Autoparametric Resonance in Mechanical Systems

When a mechanical system consists of two or more coupled vibrating components, the vibration of one of the component subsystems may destabilise the motion of the other components. This destabilisation effect is called autoparametric resonance. It is a concept that has important engineering applications. For example, flow-induced vibrations such as those caused by a rolling sea on ships, high-speed gas flows in pipelines, or turbulent air flow around aircraft wings must be considered in the design and the operation of such structures and systems.

This book is the first completely devoted to the subject of autoparametric resonance in an engineering context. With examples taken from a variety of autoparametric systems, the authors show how to carry out the first crucial step, that is, how to determine the regions of parameter space where the semitrivial solution is unstable. Using the tools of nonlinear analysis, they describe what happens in these regions and the mathematical models used to analyse them. This analysis leads to a discussion of nontrivial solutions and their stability.

The study of autoparametric systems is a lively area of current research in engineering and applied mathematics, and this book will appeal to graduate students and research workers in both disciplines.

A. Tondl has retired from the National Research Institute for Machine Design in Prague and is an Honorary Professor at the Technical University of Vienna.

M. Ruijgrok and F. Verhulst are with the Department of Mathematics at the University of Utrecht.

R. Nabergoj is a Professor of Naval Architecture at the University of Trieste.

Autoparametric Resonance in Mechanical Systems

Aleš Tondl
Thijs Ruijgrok
Ferdinand Verhulst
Radoslav Nabergoj

PUBLISHED BY THE PRESS SYNDICATE OF THE UNIVERSITY OF CAMBRIDGE
The Pitt Building, Trumpington Street, Cambridge, United Kingdom

CAMBRIDGE UNIVERSITY PRESS
The Edinburgh Building, Cambridge CB2 2RU, UK   http://www.cup.cam.ac.uk
40 West 20th Street, New York, NY 10011-4211, USA   http://www.cup.org
10 Stamford Road, Oakleigh, Melbourne 3166, Australia
Ruiz de Alarcón 13, 28014 Madrid, Spain

© Cambridge University Press 2000

This book is in copyright. Subject to statutory exception
and to the provisions of relevant collective licensing agreements,
no reproduction of any part may take place without
the written permission of Cambridge University Press.

First published 2000

Printed in the United States of America

Typeface Times Roman 11/14 pt. and Futura   *System* LaTeX 2_ε [TB]

A catalog record for this book is available from the British Library.

Library of Congress Cataloging in Publication Data
Autoparametric resonance in mechanical systems / Aleš Tondl . . . [et al.].
    p.   cm.
  Includes bibliographical references (p.   ).
  ISBN 0-521-65079-8 (hardbound)
  1. Parametric vibration. 2. Damping (Mechanics) 3. Nonlinear theories. 4. Differentiable dynamical systems. I. Tondl, Aleš.
TA355.A88   2000
620.3 – dc21   99-38309
                                                           CIP

ISBN 0 521 65079 8 hardback

Contents

Preface ix

1 Introduction 1
  1.1 What Is an Autoparametric System? 1
  1.2 Autoparametric Resonance in Practical Situations 4
  1.3 A Brief Literature Survey 7
  1.4 Models of Autoparametric Systems 8
  1.5 Scope of the Book 13

2 Basic Properties 14
  2.1 Introductory Examples 14
  2.2 A System with External Excitation 16
    2.2.1 The Semitrivial Solution and Its Stability 16
    2.2.2 Excitation-Oriented Approach 17
    2.2.3 Response-Oriented Approach 18
    2.2.4 Nontrivial Solution 20
  2.3 A Parametrically Excited System 23
    2.3.1 The Semitrivial Solution and Its Stability 23
    2.3.2 Nontrivial Solution 25
  2.4 A Self-Excited System 27
    2.4.1 The Semitrivial Solution and Its Stability 27

2.4.2 Nontrivial Solution 28
2.5 Concluding Remarks 29

3 Elementary Discussion of Single-Mass Systems 31
3.1 Introduction 31
3.2 A Primary System with External Excitation 33
    3.2.1 Semitrivial Solution and Stability 33
    3.2.2 Domains of Stability in Parameter Space 35
3.3 A Primary System with Self-Excitation 37
    3.3.1 Nontrivial Solutions in the Case of a Self-Excited Primary System 40

4 Mass–Spring–Pendulum Systems 44
4.1 The Resonantly Driven Mass–Spring–Pendulum System 44
    4.1.1 Equations of Motion 46
    4.1.2 Various Classes of Solutions 46
    4.1.3 Stability of the Semitrivial Solution 48
4.2 Nontrivial Solutions 48
    4.2.1 Quenching and Its Relation to Pendulum Damping 50
    4.2.2 Bifurcations and Stability of the Nontrivial Solutions 51
4.3 A Strongly Quenched Solution 53
    4.3.1 The Rescaled Equations 53
    4.3.2 A Quasi-Degenerate Hopf Bifurcation 55
4.4 Large-Scale Motion of the Pendulum 56
    4.4.1 The Semitrivial Solution $\varphi = \pi$ and Its Stability 57
    4.4.2 Nontrivial Solutions 57
4.5 Conclusions 61

5 Models with More Pendulums 64
5.1 Introduction 64
5.2 Formulation of the Problem 65
5.3 Stability of the Semitrivial Solution 67
5.4 An Illustrative Example 68
5.5 Conclusions 71

6 Ship Models 73
6.1 Introduction 73

6.2 Simplified Model for Heave–Roll Motion	75
6.3 Effect of the Spring Nonlinearity	79
6.4 Extension to Heave–Pitch–Roll Motion	84
6.5 Conclusions	89

7 Flow-Induced Vibrations — 90

7.1 Introduction	90
7.2 The Critical Velocity Model	93
7.3 Nontrivial Solution of the Critical Velocity Model	95
7.4 The Critical Velocity Model with Dry Friction	99
7.4.1 Stability of the Trivial Solution	100
7.4.2 The Semitrivial Solution	101
7.4.3 Nontrivial Solutions	102
7.4.4 Results of the Numerical Investigation	103
7.4.5 Remarks	106
7.5 The Vortex Shedding Model	106
7.6 Nontrivial Solution of the Vortex Shedding Model	110
7.7 Generalisation of the Critical Velocity Model	111
7.7.1 The Equations and Their Normal Form	112
7.7.2 Stability of the Semitrivial Solution	113
7.7.3 Reduction of the Normal Form	114
7.7.4 A Fixed Point with One Zero and a Pair of Imaginary Eigenvalues	116
7.7.5 Normal Form of the Equation	118
7.7.6 Bifurcations in the Normal Form	119
7.7.7 The Possibility of Šilnikov Bifurcations	125

8 Rotor Dynamics — 129

8.1 Introduction	129
8.2 The Model with Elastic Supports	130
8.3 The Linear System	134
8.4 Stability of the Semitrivial Solution	135
8.5 Nontrivial Solutions: Hysteresis and Phase Locking	139
8.6 Parametrically Forced Oscillators in Sum–Resonance	143
8.6.1 Introduction	143
8.6.2 The General Model	144
8.6.3 The Normalised Equation	146
8.6.4 Analysis of the Family of Matrices $A(\delta, \mu)$	148

9 Mathematical Methods and Ideas — 151
9.1 Basic Averaging Results — 152
9.2 Lagrange Standard Forms — 153
9.3 An Example to Illustrate Averaging — 154
9.4 The Poincaré–Lindstedt Method — 157
9.5 The Mathieu Equation with Viscous Damping — 160
9.6 The Method of Harmonic Balance — 162
9.7 Introduction to Normal Forms — 165
9.8 Normalisation of Time-Dependent Vector Fields — 169
9.9 Bifurcations — 171
    9.9.1 Local Bifurcations — 172
    9.9.2 Global Bifurcations — 173
9.10 Bifurcations in a Nonlinear Mathieu Equation — 175
    9.10.1 Normal-Form Equations — 176
    9.10.2 Dynamics and Bifurcations of the Symmetric System — 177
    9.10.3 Bifurcations in the General Case — 179
    9.10.4 Discussion — 176
9.11 The Mathieu Equation with Nonlinear Damping — 181

Bibliography — 189
Index — 195

Preface

'Usually an equation was considered to be solved if the solution had been expressed in a finite number of known functions. But this is only possible in one case out of hundred. What we always could do, or better, should do, is to solve the problem qualitatively. This means trying to find the general form of the curve which tracks the unknown function.'

Henri Poincaré
Science et Méthode

Henri Poincaré, who was a mining engineer by education and a mathematician by profession, knew very well the requirements of engineering. Quite often, in a complicated engineering problem, the researcher is interested in only one specific aspect or wants to adjust the parameters of the problem to avoid a certain phenomenon (usually one that the mathematician would find particularly interesting!). This results in an engineering practice in which the engineer derives from mathematics mainly algorithms, in particular fast techniques, to obtain numerical solutions. The mathematician, on the other hand, generally considers the engineer as a client who makes many calculations to solve useful, but mathematically not very interesting, problems.

Preface

Of course, both these views are wrong: Engineering contains, mathematically speaking, fascinating and unique problems whereas mathematics, during the past few decades, has developed a wealth of new techniques and ideas that are relevant for engineering. In this sense Poincaré's advice to obtain numerical results and at the same time qualitative insight is still extremely important. This has been one of the reasons we started our cooperation – one engineer, one physicist, and two mathematicians – in a particular field of engineering: autoparametric systems. This is a field with many applications in which the basic stability analysis is linear, although far from simple. In the case of instability, which cannot be avoided in many problems, the nonlinear analysis turns out to be very rich in phenomena, corresponding with results obtained in modern dynamical systems theory. Phenomena also arise, for instance regarding certain hidden symmetries in the engineering problem, that trigger new mathematics research.

The monograph at hand constitutes a first inventory of problems important from both engineering and mathematical points of view. We are convinced that it can be an inspiration for a considerable amount of new research in both fields.

September 1998, Prague, Utrecht, and Trieste,
A. Tondl, M. Ruijgrok, F. Verhulst, and R. Nabergoj

Chapter 1

Introduction

1.1 What Is an Autoparametric System?

The classical example of an autoparametric system is the elastic pendulum, which consists of a spring fixed at one end. The spring may swing in a plane like a pendulum and oscillate at the same time; see Figure 1.1. An interesting phenomenon arises if the ratio of the linear frequencies in the longitudinal and the transversal directions is 2:1. In that case, if we start with an oscillation of the spring in the (near) vertical direction, this normal-mode motion is unstable and energy is transferred gradually to the swinging motion and back; for a detailed description see van der Burgh (1968) or Nayfeh and Mook (1979).

This phenomenon of destabilisation of a normal mode is called autoparametric resonance or autoparametric instability. In the example of the elastic pendulum, it takes place in a conservative setting. In this book we study autoparametric resonance in an engineering context, which usually involves damping and an external source of energy.

Another example is given by a damped system of a forced, elastically mounted mass with a pendulum attached; see Figure 1.2. This system can be seen as an extension of the classical parametrically excited pendulum, and it is one of the systems that is studied extensively in various chapters of this book. To begin with, however, we discuss a general characterisation of autoparametric systems.

Introduction

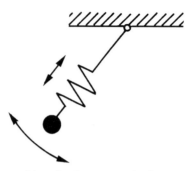

Figure 1.1: Elastic pendulum, the classic example of an autoparametric system.

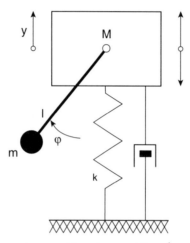

Figure 1.2: Example of an autoparametric system consisting of a mass mounted on a spring (primary system) and of a pendulum attached to the mass (secondary system).

Autoparametric systems are vibrating systems that consist of at least two constituting subsystems. One is a primary system that will generally be in a vibrating state. This primary system can be externally forced, self-excited, parametrically excited, or a combination of these. For instance, in Figure 1.2 the primary system consists of a mass mounted on a spring with damping and external periodic forcing. The primary system in this case has 1 degree of freedom. In general, the primary system has N degrees of freedom and is described by the coordinates $x_i, \dot{x}_i, i = 1, 2, \ldots, N$. The second constituting subsystem is called the secondary system. The secondary system is coupled to the primary system in a nonlinear way, but such that the secondary system can be at rest while the primary system is vibrating. We call this state the semitrivial solution. In physics it is called a normal mode. In fact, there can be an infinite number of semitrivial solutions, for instance, all the transient states to a periodic solution of the primary system, but in discussing semitrivial solutions we ignore transient states. In Figure 1.2, the secondary system consists of a pendulum attached to the mass of the primary system. Because the spring is mounted vertically, we have as the semitrivial solution that the mass is in forced periodic vibration while the pendulum is hanging vertically at rest. In general, the secondary system has n degrees of freedom and is described by the coordinates $y_j, \dot{y}_j, j = 1, 2, \ldots, n$. We can now characterise autoparametric systems as follows:

1. An autoparametric system consists of at least a primary system (coordinates $x, \dot{x}$) and a secondary system (coordinates $y, \dot{y}$) that are coupled.
2. An autoparametric system admits a semitrivial solution, which has the property $\sum_{i=1}^{N}[x_i^2(t)+\dot{x}_i^2(t)] \neq 0$, $y_i(t) = \dot{y}_i(t) = 0$, $i = 1, \ldots, n$.
3. A semitrivial solution can become unstable in certain intervals of the frequency of the excitation. This interval is known as the instability interval. It can sometimes also be defined in the case in which the primary system is self-excited.
4. In (or near) the instability intervals of the semitrivial solution we have autoparametric resonance. The vibrations of the primary system act as parametric excitation of the secondary system, which will no longer remain at rest. In this context this is called autoparametric excitation.

Autoparametric systems display many phenomena, of which we mention the following:

1. The vibrations of an autoparametric system in resonance can be periodic, quasi periodic, nonperiodic, or even chaotic.
2. The parameter interval of autoparametric resonance resulting in stable vibrations of the secondary system can be larger than the instability interval of the semitrivial solution. In such a case we have the possibility of a stable semitrivial solution coexistent with a stable

autoparametric resonant solution. They will have their own domains of attraction.
3. Saturation effects occur quite commonly. They are characterised as follows. Putting more energy in the primary system results in a strong increase of the deflections of the secondary system, whereas the increase of the vibration amplitudes of the primary system is much smaller. This phenomenon has been used to obtain absorbers. There are, however, also systems in which saturation leads to undesirable results. This will be discussed in Chapter 6 on ship models.

It is clear that in studying autoparametric systems, the determination of stability and instability conditions of the semitrivial solution is always the first step. We shall carry out this step for the models of autoparametric systems formulated in this book. In a number of cases we shall discuss what happens to the solutions if autoparametric resonance is initiated. The reader should realise that in the resonance case there are still many open problems.

1.2 Autoparametric Resonance in Practical Situations

In actual engineering problems, the loss of stability of the semitrivial response of the primary system depends on frequency tuning of the various components of the system and on the interaction (the coupling) between the primary system and the secondary system. Autoparametric vibrations occur in only a limited region of the system parameters. This property is of course of great importance for engineering purposes.

As was mentioned in Section 1.1, a saturation effect can occur in autoparametric systems. As an example of a situation for which this effect is undesirable, we mention the model of a vessel in a longitudinal sea (see also Chapter 6). The heave and pitch motions can, under certain conditions, initiate a violent rolling motion that represents the most dangerous oscillation for the vessel because it can lead to capsizing. Here the saturation phenomenon has an unfavourable effect because the increasing excitation energy of the sea waves contributes mostly to the rolling motion of the ship. Only changing the heading of the vessel with respect to the wave propagation can help to avoid the potentially catastrophic rolling motion.

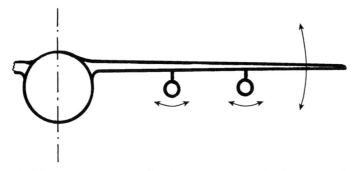

Figure 1.3: Schematic representation of airplane engines mounted under the wings by elastic suspenders.

A similar situation occurs in airplanes in which the engines are mounted under the wings by elastic suspenders; see Figure 1.3. The side deflection of the engine that is due to the elasticity of the suspenders can be described by the simple model of a mass on a leaf spring, as shown in Figure 1.4. Vertical vibrations of the wing can, under certain conditions, initiate the swinging motion of the suspended engines as marked by the arrows in the figure. As in the case of the ship model, the saturation effect, with the motion of the wings supplying the energy, can lead to violent vibrations of the engines, resulting in failure of the suspenders, i.e., a crash.

Some mechanical systems that are externally excited can be turned into autoparametric systems by use of additional subsystems, for example, by the addition of a pendulum. This has the effect of transferring the excitation energy to the added subsystem, thereby diminishing the vibration of the original basic system. Such a pendulum, or a similarily acting additional subsystem, has the same purpose as a tuned absorber commonly used in externally excited systems and self-excited systems. The use of the pendulum as an additional subsystem is well known and frequently adopted in externally excited systems. However, up until now it was not used for other types of excitation, in particular for self-excited systems. There are certain structures, machines, and devices excited by different types of self-excitation (relative dry friction, a flowing medium, etc.) for which these tools could be applied. The additional subsystem (the secondary system) need not have the form of a classical pendulum (Figure 1.2) or of a mass mounted at the end of a leaf spring (Figure 1.4). The pendulum can, for instance, be realised as a ring. As an

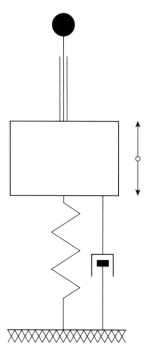

Figure 1.4: Example of an autoparametric system in which a mass mounted on the end of a long elastic element (secondary system) is coupled to a forced, damped spring (primary system).

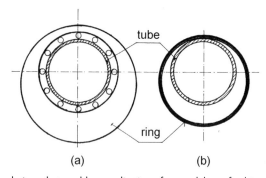

Figure 1.5: Two designs that enable a realisation of a pendulum of arbitrary short reduced length: (a) a free ring, (b) an eccentric ring mounted on a ball bearing.

example, in Figure 1.5 two different designs are shown: an eccentric ring and a free ring mounted on a ball bearing. In this way, it is possible to realise an arbitrarily reduced length of the pendulum, i.e., an arbitrary natural frequency of the secondary system. A pendulum of this type might be used in the case in which the self-excited system has the form of a rod or a tube with a circular cross section.

Note that the examples given here belong to the class of mechanical systems. It would be instructive and useful to look for examples in other fields, for example, in electrical and electronic systems that are governed by the same type of differential equations. These systems could lead to new possibilities for the application of autoparametric resonance.

1.3 A Brief Literature Survey

Until recently, research in autoparametric systems was concerned mainly with the system consisting of a vibrating single mass with an attached pendulum (Figure 1.2); see, for instance, the books by Nayfeh and Mook (1979), Schmidt and Tondl (1986), Tondl (1991b), and Cartmell (1990). In the latter, a comprehensive list of references is presented in which especially the papers of the following authors represent important contributions to this analysis: Barr, Bax, Cartmell, Ibrahim, and Roberts. The last authors have directly investigated autoparametric resonance of the system of Figure 1.2 when it is tuned into internal resonance, i.e., for the case in which the natural frequency of the harmonically excited mass–spring subsystem is approximately twice the natural frequency of the pendulum and the excitation frequency is close to the natural frequency of the excited subsystem.

The stability investigation of the semitrivial solution is presented in Tondl and Nabergoj (1990) and Tondl (1991a, 1992a, 1992b). In these papers all possible occurrences of autoparametric resonances have been determined.

A systematic study of the nontrivial solutions arising from autoparametric resonance leads into the nonlinear regime and shows many bifurcation phenomena; see Bajaj et al. (1994), Ruijgrok (1995), Ruijgrok and Verhulst (1996), and Banerjee et al. (1993, 1996).

In the literature up until 1992, the excitation of the primary system was considered to be external. In Tondl (1992b) it is shown that not only external excitation but also parametric or self-excitation of the primary system can be the source of autoparametric excitation for the secondary system when certain conditions are met. As an example we mention Thomsen's (1992) analysis of vibrations in nonshallow arches.

Mechanical models of ship behaviour in longitudinal seas, developed to investigate the possibility of rolling-motion instability, were analysed in Tondl and Nabergoj (1990, 1992) and Nabergoj and Tondl (1994); some of these results are presented in Chapter 6.

The analysis of the stability of the semitrivial solution of more complicated models with a pendulum is given in Svoboda et al. (1994), Tondl and Nabergoj (1994, 1995), and Banerjee et al. (1996).

The inclusion of other types of primary systems in the autoparametric system, such as parametrically excited and self-excited oscillators, is of practical interest and broadens the scope of autoparametric phenomena considerably. This broadening triggers many phenomena that are of interest both mathematically and from the point of view of mechanics. A comprehensive review of physical models that exhibit autoparametric resonance is given in Tondl and Nabergoj (1993). The analysis of the models presented in this book involves mathematical concepts and methods, summarised in Chapter 9 of this book. More extensive introductions to the modern analysis of nonlinear vibrations are given in Thompson and Stewart (1986), Thomsen (1997), Moon (1987), Guckenheimer and Holmes (1983), and Dankowicz (1997).

1.4 Models of Autoparametric Systems

We have seen one example of an autoparametric system that has 2 degrees of freedom (see Figure 1.2). It consists of a mass mounted on a spring (primary system) and of a pendulum (secondary system) attached to the mass. The mass on the spring can move in the vertical direction only, and it is periodically excited. The forces acting on the pendulum result from gravity and from the spring. This system is analysed in Chapter 4.

Another interesting set of phenomena arises if we admit rotation of the pendulum. It turns out that there are various parameter regimes with nonperiodic and chaotic motion; see Section 4.4.

A related simple model consists of a mass, supported by a spring that is mounted on the end of a long elastic element, for instance, a leaf spring; see Figure 1.4. This is also a simple model for a vertical, oblong body that is mounted on a coherent elastic foundation. Again the excited spring with mass is the primary system; the leaf spring–mass represents the secondary system.

Models of Autoparametric Systems

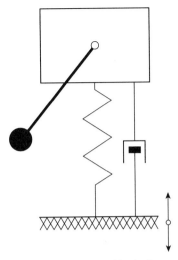

Figure 1.6: The simplest model of a ship excited by the longitudinal waves of the sea.

The excitation of the primary system in the first example need not act on the mass. It can act kinematically because of the periodic motion of the spring support (see Figure 1.6). This is a simple model for the motion of a vessel that is due to the longitudinal waves of the sea. This system is analysed in Chapter 6 along with more complicated ship models.

A natural extension of the first model arises when a chain of masses and springs, mounted vertically, is introduced for the primary system. To one of the masses a pendulum is attached, representing the excited system (see Figure 1.7).

An example in which the primary system is excited parametrically is shown in Figure 1.8. This represents a system in which the mass of the primary system is mounted vertically between two nonlinear springs whose supports are moving periodically. These periodic motions have the same amplitude and frequency and are in fact identical, but they cause the supports to move in opposite directions because of a phase shift. If the springs also obey certain conditions for their characteristics, we have pure parametric excitation of the primary system. A pendulum attached to the mass again represents the secondary system. Such a system is studied in Chapters 4 and 6 in which the motion of the supports is taken to be harmonic.

An important class of engineering problems consists of vibrations induced by cross flow, which represent examples for which the primary

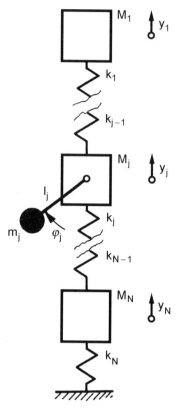

Figure 1.7: Example of an autoparametric system consisting of a chain of masses and springs (primary system) in which a pendulum is attached to some of the masses (secondary system).

system is self-excited. A simple model is a spring–pendulum system, as depicted in Figure 1.9, which is excited by cross flow with constant flow velocity. Vibrations involving cross flow are discussed in Chapter 7.

A special group of autoparametric systems is formed by rotors with a vertical shaft that is excited kinematically through an elastic support. An example is shown in Figure 1.10, in which a rigid rotor is mounted elastically in both the radial and the axial directions and the axial thrust bearing is considered to be a joint. The excitation arises, for instance, by the vibrations of nearby machinery. Autoparametric rotor systems are considered in Chapter 8.

Another group of models that show autoparametric resonance is formed by single-mass systems with 2 degrees of freedom, for instance, a mass moving in a horizontal or a vertical plane. If the system is excited

Models of Autoparametric Systems

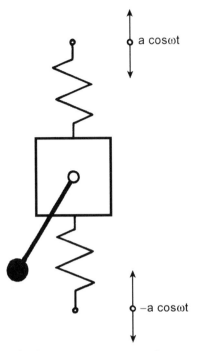

Figure 1.8: Example of autoparametric system with parametric excitation.

Figure 1.9: Example of an autoparametric system excited by flow consisting of a single mass on a spring, to which a pendulum is attached, moving in one direction. The flow moves the mass and the spring (primary system) but not the pendulum (secondary system).

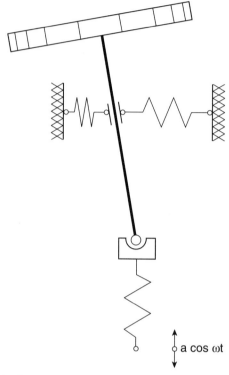

Figure 1.10: Example of autoparametric system in which a rotor system with a vertical shaft is kinematically excited through an elastic support element.

externally, parametrically, or by self-excitation, in one direction only, and if it satisfies moreover the conditions of Section 1.1, we again have an autoparametric system. Such problems are considered in Chapter 3. Other, quite different types of problems arise in dealing with nonlinear wave equations. In such problems one often applies a Galerkin projection or a finite Fourier mode expansion that results in a finite chain of oscillators that is proposed as an approximating system of the original nonlinear wave problem. The question is then whether higher-order modes, which have been left out in the Galerkin projection, can be excited by lower-order modes. This would result in a transfer of energy from lower-order to higher-order modes, which destroys, after some time, the approximate character of the Galerkin projection. In a number of cases this can be interpreted as autoparametric resonance and can be described by one of the models in this book.

1.5 Scope of the Book

This book should be considered as partly a literature survey and partly a workbook. It is a text in which a large amount of various autoparametric systems, corresponding with very different engineering contexts, is studied. For these systems we carry out the first, crucial step, which is the determination of the sets in parameter space in which the semitrivial solution is unstable. These are the regions of autoparametric resonance. To find out what happens in these resonance regions and to further analyse the mathematical models of autoparametric systems, we use the tools of nonlinear dynamics. This enables us to find nontrivial solutions, determine their stability, analyse branching of solutions, and study the various domains of attraction in the case of coexisting stable solutions. In some of the cases discussed in this book, we present a more detailed mathematical analysis. For instance, in Chapter 7 we show that, under certain conditions, chaotic solutions can be expected in a model with a self-excited primary system. In other cases we rely on numerical results. An example can be found in Chapter 4, in which we consider a model consisting of a rotating pendulum (the secondary system) attached to a spring that oscillates vertically (the primary system). Finally, many of the models presented in this book are the subject of current research, and a more detailed analysis in those cases is not yet available.

Autoparametric systems show a, perhaps surprising, amount of new phenomena and challenges to both engineers and mathematicians. We hope that this book will be an inspiration for future research into this subject.

Chapter 2

Basic Properties

2.1 Introductory Examples

The examples presented in this chapter are artificial in the sense that they were not proposed to model real-life problems, but to analyse and demonstrate the basic properties of autoparametric systems with examples as simple as possible. Note that for these systems we need at least 2 degrees of freedom and a nonlinear, resonant interaction. The examples can be viewed as realistic when we consider them as modeling single-mass systems with 2 degrees of freedom. We return to this point of view in some elementary examples discussed in Chapter 3.

We consider three examples, each consisting of two subsystems that have 1 degree of freedom. Each example contains a different kind of primary system, characterised subsequently by external excitation, parametric excitation, and self-excitation. The first system is characterised by an externally excited primary system. The governing equations, transformed into dimensionless form, are

$$x'' + \kappa_1 x' + x + \gamma_1 y^2 = a\eta^2 \cos \eta \tau,$$
$$y'' + \kappa_2 y' + q^2 y + \gamma_2 xy = 0, \qquad (2.1.1)$$

where $\kappa_1 > 0$ and $\kappa_2 > 0$ are the damping coefficients, γ_1 and γ_2 are the nonlinear coupling coefficients, $q = \omega_2/\omega_1$ is the tuning coefficient that expresses the ratio of natural frequencies of the undamped linearised secondary system and the primary system, $a\eta^2$ expresses the amplitude of the external excitation, and $\eta = \omega/\omega_1$ is the forcing frequency. Here, ω is the dimensional frequency of the excitation and ω_1 and ω_2 are the natural frequencies of the primary system and the secondary system, respectively.

The second system is an example of a parametrically excited primary system that is governed by the following equations of motion:

$$x'' + \kappa_1 x' + (1 + a\cos 2\eta\tau)x + \gamma_1(x^2 + y^2)x = 0,$$
$$y'' + \kappa_2 y' + q^2 y + \gamma_2 xy = 0, \quad (2.1.2)$$

where κ_1, κ_2, γ_1, γ_2, and q are coefficients similar to those for the first system, η is the excitation frequency, and a is the coefficient of the parametric-excitation term.

The third system contains a self-excited primary system. This system is governed by the following equations of motion:

$$x'' - (\beta - \delta x'^2)x' + x + \gamma_1 y^2 = 0,$$
$$y'' + \kappa y' + q^2 y + \gamma_2 xy = 0, \quad (2.1.3)$$

where κ, γ_1, γ_2, and q have meanings similar to those of the preceding cases and $\beta > 0$ and $\delta > 0$ are the coefficients of the terms representing the self-excitation of Rayleigh type.

In the first and the third examples we have chosen the term $\gamma_1 y^2$ as the nonlinear coupling term in the equation for the primary system. This choice was made because $\gamma_1 y^2$ is the lowest-order term that produces a resonant interaction when $\omega_1 : \omega_2 = 2 : 1$. It is precisely this resonance that is studied.

In the second example, however, restricting the coupling term to $\gamma_1 y^2$ would not lead to a bounded semitrivial solution. We have therefore chosen $\gamma_1(x^2 + y^2)x$ as the coupling term in this example. Such a term might arise, for instance, when the underlying system is symmetric under $x \to -x$. This situation occurs in the single-mass system studied in Chapter 3.

The coupling term $\gamma_2 xy$ in the secondary system is the same for the three alternatives, again for the sake of simplicity. In each system we obtain the semitrivial solution by putting $y = 0$, $y' = 0$. The choice

Basic Properties

of the coupling terms affects the type of autoparametric resonance that occurs in the system. We discuss this problem in Section 2.5.

2.2 A System with External Excitation

2.2.1 The Semitrivial Solution and Its Stability

To find the semitrivial solution of Eq. (2.1.1) we put

$$x(\tau) = R\cos(\eta\tau + \psi_1), \quad y(\tau) = 0. \qquad (2.2.1)$$

This yields the solution for R:

$$R = R_0 = \frac{a\eta^2}{\Delta^{1/2}}, \quad \Delta = (1-\eta^2)^2 + \kappa_1^2\eta^2. \qquad (2.2.2)$$

Note that when $\kappa_1 = \mathcal{O}(\varepsilon)$ and $\eta = 1 + \mathcal{O}(\varepsilon)$, the amplitude of the semitrivial solution is $R_0 = \mathcal{O}(a/\varepsilon)$. This situation is related to the main resonance for the primary system, and it will be one of the cases under consideration.

The stability investigation of the semitrivial solution will show the intervals of the excitation frequency where this semitrivial solution is unstable and a nontrivial solution will arise. Inserting the expressions

$$x = R_0 \cos(\eta\tau + \psi_1) + u, \quad y = 0 + v,$$

into Eqs. (2.1.1) then yields, in linear approximation,

$$u'' + \kappa_1 u' + u = 0,$$
$$v'' + \kappa_2 v' + [q^2 + \gamma_2 R_0 \cos(\eta\tau + \psi_1)]v = 0. \qquad (2.2.3)$$

The solution $u = 0$ of the first equation of Eqs. (2.2.3) is asymptotically stable. Thus the second equation of Eqs. (2.2.3) fully determines the stability of the semitrivial solution. This equation is of Mathieu type, and its main instability domain is found for values of q near $\frac{1}{2}\eta$. The Mathieu equation is discussed in fuller detail in Chapter 9.

We assume κ_2 and γ_2 to be small, and we write

$$\kappa_2 = \varepsilon\hat{\kappa}_2, \quad \gamma_2 = \varepsilon\hat{\gamma}_2, \quad q^2 = \tfrac{1}{4}\eta^2 + \varepsilon\sigma_2.$$

Putting $v_1 = v$, $v_2 = v'$ and translating the time variable so that $\psi_1 = 0$

gives the equations

$$v_1' = v_2,$$
$$v_2' = -\tfrac{1}{4}\eta^2 v_1 - \varepsilon(\kappa_2 v_2 + \sigma_2 v_1 + \gamma_2 R_0 \cos\eta\tau\, v_1), \quad (2.2.4)$$

where it is assumed that $R_0 = \mathcal{O}(1)$ as $\varepsilon \to 0$ and the hats have been dropped. We subsequently leave out the limit $\varepsilon \to 0$, as it is assumed that ε is always a small parameter.

As in Section 9.5, the boundary of the main instability domain can be found by use of the averaging method. We find to first order in ε that

$$\sigma_2^2 + \tfrac{1}{4}\kappa_2^2\eta^2 - \tfrac{1}{4}\gamma_2^2 R_0^2 = 0. \quad (2.2.5)$$

Two different approaches on how to use condition (2.2.5) can be applied. The first one of these can be called the excitation-oriented approach and the second the response-oriented approach. We describe these two methods in Subsections 2.2.2 and 2.2.3, respectively.

2.2.2 Excitation-Oriented Approach

In the first approach, the expressions for R_0 from Eqs. (2.2.2) are inserted into Eq. (2.2.5). This yields the critical value a_c for the excitation amplitude:

$$a_c = \frac{\Delta^{1/2}}{\gamma_2 \eta^2} \left(\sigma_2^2 + \kappa_2^2\eta^2\right)^{1/2}. \quad (2.2.6)$$

For values of a above this critical value the semitrivial solution is unstable. In particular, from Eq. (2.2.6) it follows that when $\eta = 1 + \mathcal{O}(\varepsilon)$ then $\Delta^{1/2} = \mathcal{O}(\varepsilon)$ and so also $a_c = \mathcal{O}(\varepsilon)$.

As an example, in Figure 2.1 the instability threshhold $\gamma_2 a_c$ of the semitrivial solution is shown. Note that we have multiplied the amplitude of excitation a by the coefficient of nonlinearity γ_2. Also, to obtain a more convenient representation, the direction of the $\gamma_2 a$ axis has been changed so that minima appear as maxima and the instability domain now lies below the surface. The system parameters κ_1 and κ_2 are given in the diagrams directly. Moreover, in the (η, q) plane the lines $\eta = 1$ and $\eta = 2q$ are marked. The figure shows that close to $\eta = 1$ and $\eta = 2q$ the instability threshhold exhibits local minima.

Basic Properties

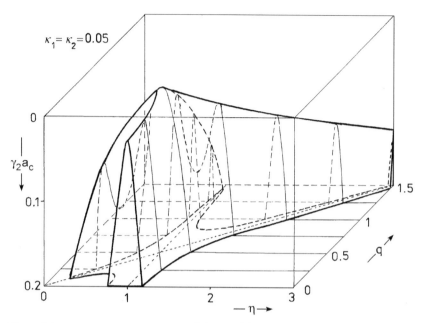

Figure 2.1: Axonometric representation of the instability threshhold $\gamma_2 a_c$ of the semitrivial solution. The instability region is below the surface. The values of the parameters are $\kappa_1 = \kappa_2 = 0.05$.

2.2.3 Response-Oriented Approach

In this approach, we use the amplitude of the response rather than the amplitude of the excitation to characterise the stability of the semitrivial solution. In many applications this is the preferred method. The amplitude of the response is R_0, so from Eq. (2.2.5) it follows that the critical value of R_0 (i.e., where the semitrivial solution loses stability) is given by

$$R_0 = R_c(\eta) = \frac{1}{\gamma_2}\left(\sigma_2^2 + \kappa_2^2\eta^2\right)^{1/2}. \qquad (2.2.7)$$

Plotting $R_c(\eta)$ together with the amplitude $R_0(\eta)$ of the semitrivial solution in an amplitude-frequency diagram gives the values of the frequency η for which the semitrivial solution is unstable.

This is demonstrated in Figure 2.2 by an example with the following parameter values: $\kappa_1 = 0.10$, $\kappa_2 = 0.05$, $\gamma_2 = 0.10$, and $q = 0.75$. The

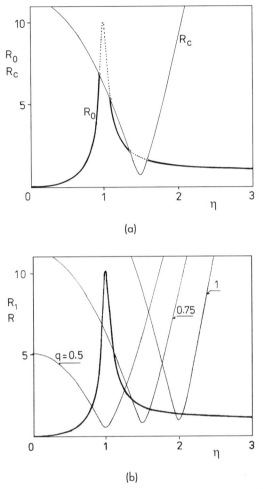

Figure 2.2: Vibration amplitude curve R_0 corresponding to the semitrivial solution (stable solution, heavy solid curves; unstable solution, dotted curves) and the stability boundary curve R_c (light solid curves) as functions of the excitation frequency η. The following values have been used: $\kappa_1 = 0.10, \kappa_2 = 0.05, \gamma_2 = 0.10, q = 0.75$.

figure shows the frequency response curve $R_0(\eta)$ as well as $R_c(\eta)$, the latter marked by a solid light curve. Parts of the curve to which unstable solutions correspond are indicated by dotted curves. As we can see, there exist unstable parts in the response, and these are located near $\eta = 1$ and $\eta = 2q$, in accordance with the preceding analysis.

2.2.4 Nontrivial Solution

We now look for a nontrivial periodic solution in the case in which $q \approx \frac{1}{2}\eta$. As was noted in Subsection 2.2.3, we must then take $a = \mathcal{O}(\varepsilon)$. Rescaling Eqs. (2.1.1) through

$$a = \varepsilon \hat{a}, \quad \kappa_1 = \varepsilon \hat{\kappa}_1, \quad \gamma_1 = \varepsilon \hat{\gamma}_1,$$
$$\eta^2 = 1 - \varepsilon \sigma_1, \quad q^2 = \tfrac{1}{4}\eta^2 + \varepsilon \sigma_2,$$

gives the following equations (with the hats dropped):

$$x'' + \eta^2 x = -\varepsilon(\kappa_1 x' + \sigma_1 x + \gamma_1 y^2 - a\eta^2 \cos \eta \tau),$$
$$y'' + \tfrac{1}{4}\eta^2 y = -\varepsilon(\kappa_2 y' + \sigma_2 y + \gamma_2 xy). \qquad (2.2.8)$$

For $\varepsilon = 0$, the solutions to Eqs. (2.2.8) can be written as

$$x = R_1 \cos(\eta \tau + \psi_1), \quad y = R_2 \cos\left(\tfrac{1}{2}\eta \tau + \psi_2\right).$$

A $4\pi/\eta$-periodic solution can be found with the Poincaré–Lindstedt method (see Chapter 9). This method leads to the following system of conditions, up to $\mathcal{O}(\varepsilon)$:

$$\sigma_1 R_1 + \tfrac{1}{2}\gamma_1 R_2^2 \cos(\psi_1 - 2\psi_2) - a \cos \psi_1 = 0,$$
$$\kappa_1 R_1 - \tfrac{1}{2}\gamma_1 R_2^2 \sin(\psi_1 - 2\psi_2) + a \sin \psi_1 = 0,$$
$$\sigma_2 R_2 + \tfrac{1}{2}\gamma_2 R_1 R_2 \cos(\psi_1 - 2\psi_2) = 0,$$
$$-\kappa_2 R_2 - \gamma_2 R_1 R_2 \sin(\psi_1 - 2\psi_2) = 0. \qquad (2.2.9)$$

Note that $\eta = 1 + \mathcal{O}(\varepsilon)$; therefore we have replaced expressions such as $a\eta^2$ and $\kappa_1\eta$ with a and κ_1.

The vibration amplitude of the x coordinate follows from the last two equations of system (2.2.9), and the result is

$$R_1 = \frac{2}{\gamma_2}\left(\sigma_2^2 + \tfrac{1}{4}\kappa_2^2\right)^{1/2}. \qquad (2.2.10)$$

From the first two equations of system (2.2.9) the following quadratic equation for the amplitude of the y coordinate is obtained:

$$z^2 + Bz + A = 0, \quad z = \tfrac{1}{2}\gamma_1\gamma_2 R_2^2, \qquad (2.2.11)$$

with

$$A = 4(\sigma_1^2 + \kappa_1^2)(\sigma_2^2 + \kappa_2^2) - \gamma_2^2 a^2,$$
$$B = 4(\kappa_1\kappa_2 - \sigma_1\sigma_2). \qquad (2.2.12)$$

Let $D = B^2 - 4A = 16[\frac{1}{4}\gamma_2^2 a^2 - (\sigma_1\kappa_2 + \sigma_2\kappa_1)^2]$; then Eq. (2.2.11) has no solutions if $A > 0$ and $D < 0$, two solutions if $A > 0$ and $D > 0$, and one solution if $A < 0$, irrespective of the value of D. Note that the condition $A < 0$ is equivalent to

$$a > \frac{2}{\gamma_2}(\sigma_1^2 + \kappa_1^2)^{1/2}(\sigma_2^2 + \kappa_2^2)^{1/2} = a_c. \qquad (2.2.13)$$

In other words, $A < 0$ is equivalent to the condition that the semitrivial solution is unstable. So in this case system (2.1.1) has an unstable semitrivial solution and a stable periodic solution. If $A > 0$ the situation is more complicated.

In this system we see the so-called saturation phenomenon occurring. Assume that all the parameters except a are constant and such that $B > 0$. Letting a increase from 0 to a_c, we see that the stable response of the system will be the semitrivial solution. From Eqs. (2.2.2) it follows that the amplitude of this solution, which is given by

$$R_0 = \frac{a}{(\sigma_1^2 + \kappa_1^2)^{1/2}},$$

grows linearly with a. At $a = a_c$ the semitrivial solution loses stability in a supercritical period-doubling bifurcation. The amplitude of the (semitrivial) response is then

$$R_0 = \frac{a_c}{(\sigma_1^2 + \kappa_1^2)^{1/2}} = \frac{2}{\gamma_2}(\sigma_1^2 + \kappa_1^2)^{1/2}.$$

When $a > a_c$, it follows from Eq. (2.2.10) that the x component of the response remains constant when a is increased. The y component, which can be calculated by the solution of Eq. (2.2.11), grows with increasing a. Thus, when the excitation amplitude is increased, the portion of the energy supplied by the external source to the primary system remains constant and the whole increment of energy flows to the excited subsystem.

Basic Properties

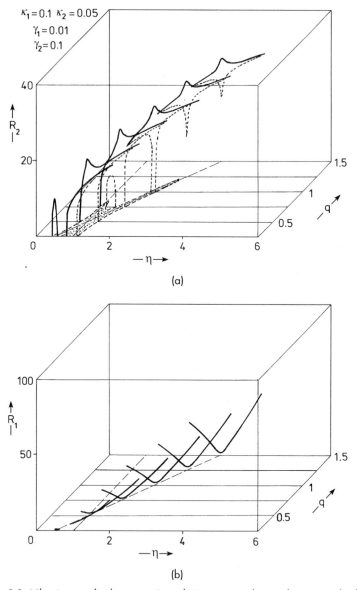

Figure 2.3: Vibration amplitude curves R_1 and R_2 corresponding to the nontrivial solution as functions of the excitation frequency η and tuning ratio q. Stable solutions are marked by solid curves and unstable solutions by dotted curves. The following values have been used: $\kappa_1 = 0.1, \kappa_2 = 0.05, \gamma_1 = 0.01, \gamma_2 = 0.10$.

In Figures 2.3 we have plotted the values of R_1 and R_2 as functions of q and η for specific values of the parameters κ_1, κ_2, γ_1, and γ_2. These curves are arranged in axonometric view, and the values are marked directly in the diagrams. Parts of the curves corresponding to stable solutions are marked by heavy solid curves. The unstable solutions are marked by dashed lines. In the (η, q) plane bias, the scale lines for certain constant values of q are marked by light solid straight lines interrupted in those intervals of η where, for certain values of the excitation frequency, only one solution of R_2 for the nontrivial solution exists. This area in the (η, q) plane is dotted and its boundary is marked by a dashed curve. These diagrams show that the domain of existence of the stable nontrivial solution is broader than that of the semitrivial solution instability. It follows that there exist frequency intervals where two locally stable periodic solutions exist: both the semitrivial solution and a nontrivial solution (autoparametric resonance), and consequently two domains of attraction as well.

To illustrate the transient behaviour of the system when the excitation frequency η is slowly increased and subsequently decreased, the values of $R_1(\eta)$ and $R_2(\eta)$ are shown in Figure 2.4 for $q = 0.25$ and 0.75 and for the following parameter values: $\kappa_1 = 0.10$, $\kappa_2 = 0.05$, $\gamma_1 = 0.01$, and $\gamma_2 = 0.10$. It can be seen that there exists one interval of η where two locally stable solutions exist. At the boundary of this interval the character of the solution changes by a jump.

2.3 A Parametrically Excited System

2.3.1 The Semitrivial Solution and Its Stability

After the equations of motion are rescaled and the hats are dropped, the semitrivial solution of Eqs. (2.1.2) is given by $y = 0$ and x a solution of

$$x'' + \varepsilon\kappa_1 x' + (\eta^2 + \varepsilon a \cos 2\eta\tau)x + \varepsilon\sigma_1 x + \varepsilon\gamma_1 x^3 = 0. \quad (2.3.1)$$

Assuming that $x = R_0 \cos(\eta\tau + \psi_0)$, we can find equations for R_0 and ψ_0. After averaging over τ and a time scaling, these become

$$R_0' = -\kappa_1 \eta R_0 + \tfrac{1}{2} a R_0 \sin 2\psi_1,$$

$$\psi_0' = \sigma_1 + \tfrac{1}{2} a \cos 2\psi_1 + \tfrac{3}{4}\gamma_1 R_0^2. \quad (2.3.2)$$

Basic Properties

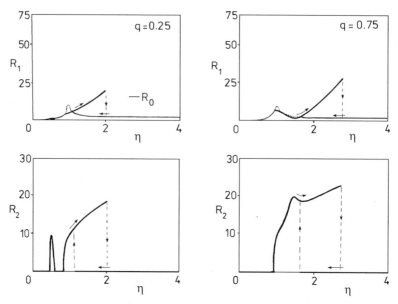

Figure 2.4: Vibration amplitude curves $R_1 = R_1(\eta)$ and $R_2 = R_2(\eta)$ corresponding to the nontrivial solution when the excitation frequency η is increased and subsequently decreased. The arrows mark the sense of changes and jumps. The following values have been used: $\kappa_1 = 0.10$, $\kappa_2 = 0.05$, $\gamma_1 = 0.01$, $\gamma_2 = 0.10$, $q = 0.25$ and 0.75.

Equilibrium solutions of Eqs. (2.3.2) correspond to $2\pi/\eta$-periodic solutions. We find that the amplitude of these periodic solutions is given by

$$R_0^2 = \frac{4}{3}\frac{1}{\gamma_1}\left[-\sigma_1 \pm \left(\tfrac{1}{4}a^2 - \kappa_1^2\eta^2\right)^{1/2}\right]. \tag{2.3.3}$$

From the averaged equations it follows that the plus sign corresponds to a stable solution of Eq. (2.3.1) and the minus sign to an unstable solution. The stability of the semitrivial solution is determined by

$$y'' + \varepsilon\kappa_2 y' + \tfrac{1}{4}\eta^2 y + \varepsilon\sigma_2 y + \varepsilon\gamma_2 r_0 \cos\eta\tau y = 0, \tag{2.3.4}$$

where r_0 is the solution of Eq. (2.3.3) corresponding to the plus sign. The boundary of the main instability domain is, to first order in ε, given by

$$\tfrac{1}{4}\gamma_2^2 r_0^2 = \sigma_2^2 + \tfrac{1}{4}\kappa_2^2\eta^2. \tag{2.3.5}$$

Note that this result is very similar to Eq. (2.2.5) in the preceding example.

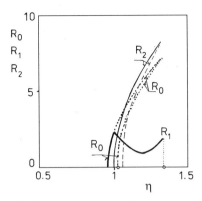

Figure 2.5: Vibration amplitude curves R_0, R_1, and R_2 corresponding to the semitrivial and the nontrivial solution as functions of the excitation frequency η. Stable solutions are marked by solid curves and unstable solutions by dashed and dotted curves. The following values have been used: $\kappa_1 = \kappa_2 = 0.075$, $\gamma_1 = 0.02$, $\gamma_2 = 0.10$, $\varepsilon = 0.20$, $q = 0.60$.

2.3.2 Nontrivial Solution

Nontrivial solutions can be found as in Subsection 2.3.1 by the introduction of

$$x = R_1 \cos(\eta\tau + \psi_1), \quad y = R_2 \cos\left(\tfrac{1}{2}\eta\tau + \psi_2\right).$$

Applying the Poincaré–Lindstedt method then yields the following set of conditions:

$$-\kappa_1 \eta R_1 + \tfrac{1}{2} a R_1 \sin 2\psi_1 = 0,$$
$$\sigma_1 + \tfrac{1}{2} a \cos 2\psi_1 + \tfrac{3}{4}\gamma_1 R_1^2 + \tfrac{1}{2}\gamma_1 R_2^2 = 0,$$
$$-\tfrac{1}{2}\kappa_2 \eta R_2 - \tfrac{1}{2}\gamma_2 R_1 R_2 \sin(\psi_1 - 2\psi_2) = 0,$$
$$\sigma_2 + \tfrac{1}{2}\gamma_2 R_1 \cos(\psi_1 - 2\psi_2) = 0. \tag{2.3.6}$$

This yields

$$\tfrac{1}{4}\gamma_2^2 R_1^2 = \sigma_2^2 + \tfrac{1}{4}\kappa_2^2 \eta^2,$$
$$\tfrac{1}{2}\gamma_1 R_2^2 = -\tfrac{3}{4}\gamma_1 R_1^2 - \sigma_1 \pm \left(\tfrac{1}{4}a^2 - \kappa_1^2 \eta^2\right)^{1/2}. \tag{2.3.7}$$

As in the preceding example, we have a saturation phenomenon. The following parameter values are taken for explicit examples: $\kappa_1 = \kappa_2 = 0.075$, $\gamma_1 = 0.02$, $\gamma_2 = 0.10$, and $\varepsilon = 0.20$. In Figure 2.5, we

Basic Properties

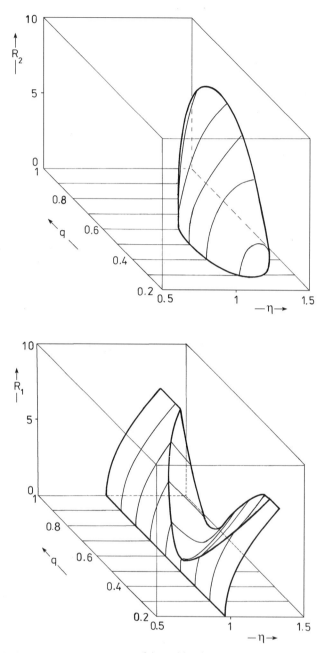

Figure 2.6: Axonometric representation of the stable vibration amplitudes R_1 and R_2 corresponding to the nontrivial solution as functions of the excitation frequency η and tuning ratio q. The following values have been used: $\kappa_1 = \kappa_2 = 0.075$, $\gamma_1 = 0.02$, $\gamma_2 = 0.10$, $\varepsilon = 0.20$.

show the amplitudes R_0, R_1, R_2 as functions of η for the case $q = 0.60$; the solid curves represent stable solutions and the dashed or dotted ones represent unstable solutions. Again there is an interval of η where both semitrivial and nontrivial solutions exist, their realisation depending on initial conditions. This instability interval lies between the points on the η axis marked by circles.

To illustrate the influence of the tuning coefficient q, Figure 2.6 shows in axonometric view the amplitudes R_1 (both semitrivial and nontrivial solutions) and R_2 as functions of η and q (only stable solutions). We can see that there exists only one interval of q where autoparametric resonance is initiated, i.e., in the neighbourhood of $q = \frac{1}{2}\eta$. The domain of autoparametric resonance occurrence is relatively narrower compared with the system that has an externally excited primary system.

2.4 A Self-Excited System

2.4.1 The Semitrivial Solution and Its Stability

In contrast with the preceding two examples, system (2.1.3) is autonomous, so the periodic solutions will not have a definite phase associated with them. After the equations of motion are rescaled and the carets are dropped, the semitrivial solution of system (2.1.3) is given by $y = 0$, $y' = 0$, and x a solution of

$$x'' - \varepsilon(\beta - \delta x'^2)x' + x = 0. \qquad (2.4.1)$$

Writing $x = R_0 \cos(\tau + \psi_0)$ and averaging over τ yields the equations

$$\begin{aligned} R_0' &= \tfrac{1}{2}\varepsilon \left(\beta R_0 - \tfrac{3}{4}\delta R_0^3\right), \\ \psi_0' &= 0. \end{aligned} \qquad (2.4.2)$$

From the averaged equations, we find a stable periodic orbit with amplitude

$$R_0 = \left(\tfrac{4}{3}\frac{\beta}{\delta}\right)^{1/2}. \qquad (2.4.3)$$

Using the same method as in the two preceding examples, we find that the semitrivial solution is stable if

$$\frac{\beta \gamma_2}{3\delta} \leq \sigma_2^2 + \tfrac{1}{4}\kappa^2. \tag{2.4.4}$$

2.4.2 Nontrivial Solution

To find nontrivial periodic solutions, we transform

$$x = R_1 \cos(\tau + \psi_1), \quad y = R_2 \cos\left(\tfrac{1}{2}\tau + \psi_2\right).$$

It was noted that the periodic solutions of system (2.1.3) will not have a definite phase. We can, however, define the phase difference between the x and the y components by $\phi = \psi_1 - 2\psi_2$. Applying the Poincaré–Lindstedt method leads to the following set of equations:

$$\beta R_1 - \tfrac{3}{4}\delta R_1^3 + \tfrac{1}{2}\gamma_1 R_2^2 \sin\phi = 0,$$
$$-\tfrac{1}{2}\kappa R_2 - \tfrac{1}{2}\gamma_2 R_1 R_2 \sin\phi = 0,$$
$$\tfrac{1}{4}\gamma_1 \frac{R_2^2}{R_1} \cos\phi - 2\sigma_2 - \gamma_2 R_1 \cos\phi = 0. \tag{2.4.5}$$

The presence of the combination angle $\phi = \psi_1 - 2\psi_2$ in the equations is related to a synchronisation phenomenon.

A long but straightforward calculation shows that for the amplitudes R_1 and R_2 we find at most two solutions. Stability analysis of these solutions indicates that, as in the preceding cases, the stability boundary of the semitrivial solution, as given by relation (2.4.4), corresponds to a period-doubling bifurcation. Other bifurcations that can occur are the saddle node, Hopf, and a codimension-two bifurcation characterised by eigenvalues $\pm i, 0$. It is well known that chaotic phenomena are associated with this latter bifurcation (see Guckenheimer and Holmes, 1983).

It is clear that the system with a self-excited primary system in a first calculation produces more complex phenomena. An extensive analysis of this system is given in Chapter 7.

2.5 Concluding Remarks

The purpose of this chapter has been to demonstrate that autoparametric resonance can be initiated by a primary system that is externally excited, parametrically excited, or self-excited. Also, we gave a first indication of what happens when the semitrivial solution becomes unstable. We have also seen that the domain of existence in parameter space of autoparametric resonance, i.e., the domain where nontrivial solutions exist, can be substantially larger than the domain of instability of the semitrivial solution.

In the examples that were discussed here, the saturation phenomenon occurs (most strongly in the cases of external and self-excitation). This phenomenon offers the possibility of using autoparametric resonance to restrict the vibration amplitudes in a part of a system where energy is being supplied.

Moreover, as we have seen in the preceding examples, nonlinear resonant interaction may produce synchronisation phenomena in the case of self-excited systems. The nontrivial solutions, i.e., the periodic solutions that emerge in certain parameter domains, may display locking of the two systems, the primary system and the secondary system.

We make another remark on the tuning of the vibrating systems and resonance. Let us focus our attention on the first example in which the primary system is externally excited with frequency η. The basic frequency of the primary system is 1; the basic frequency of the secondary system is q. In the calculations we took $q \approx \frac{1}{2}\eta$, as this invokes the most prominent Mathieu instability domain in the Mathieu equation, which describes the stability of the semitrivial solution. However, additional resonances, occurring for other discrete values of q, also correspond with interesting phenomena. We simply took the case that is the most important one from the physical point of view.

These tuning considerations involve linear resonance, but we have the additional phenomenon of nonlinear resonance, which produces synchronisation and many other features. Suppose, for instance, that in system (2.1.2) we interchange the term xy^2 in the first equation with the term xy in the second equation. The reader may verify that, to a first-order approximation, as calculated in Section 2.3, the nonlinear interaction that shapes the nontrivial solution found in this section vanishes if $\eta = 1$. However, there is a nonlinear interaction for some other values of η.

This shows that our results and the corresponding phenomena are sensitive to which nonlinearities are taken, in the sense that certain nonlinear terms generically produce nontrivial behaviour, whereas other nonlinear terms are of less importance, producing small quantitative but not qualitative changes. It should be clear that similar considerations hold for the other examples of this chapter.

It is important to have general insight into the part played by nonlinearities in our equations. In Chapter 3 we consider some other resonances and nonlinearities.

The preliminary analysis of the models in this chapter already shows a number of elementary bifurcations (such as saddle node, Hopf, and period doubling) and, in the case of the last example, a not so elementary one of codimension two.

Chapter 3

Elementary Discussion of Single-Mass Systems

3.1 Introduction

This chapter contains an elementary discussion of systems that are modelled as single-mass systems with 2 degrees of freedom. It can be seen as a natural sequel to Chapter 2 with different resonances and different nonlinearities. Single-mass systems form a natural context for the possible occurrence of autoparametric resonance. Consider, for instance, rods, beams, or tubes by replacing them with a corresponding structure with the mass concentrated in the centre of the element. Using Galerkin or Ritz methods, we assume that this structure is vibrating in one natural mode. This assumption can, for instance, be used when the excitation frequency is close to the first natural frequency or, in the case of a self-excited system, that in most cases systems are vibrating in the first mode. This holds, for instance, in cases like galloping of high-voltage lines or vibrations of tubes in cross flow; see also Chapter 7. We hope that our simplified models will enable us to understand more complicated vibration phenomena.

We formulate three different systems in which some kind of excitation in the x direction takes place. Considered as an autoparametric system, this will represent the primary system. It can, for instance, be caused by the kinematic excitation that is due to the vibrations of the frame or housing in which the rod, beam, or tube under consideration is clamped. On the other hand, vibrations may vary the boundary conditions; this adds a variation to the constraint that can be the source of parametric excitation. As mentioned previously, with more details in Chapter 7, cross flow may induce self-excited vibrations.

If we have linear systems that are uncoupled, the vibrations do not spread to other directions. However, we assume that we have nonlinear terms of the third order in the system of differential equations, coupling the 2 degrees of freedom with each other. Such nonlinear terms arise quite commonly from the effect of nonlinear boundary conditions in constraints of rods and beams. The second degree of freedom, in the y direction, will represent the secondary system. We formulate three systems like this with the purpose of studying basic properties for two of them.

For reasons of comparison, the same cubic, nonlinear coupling term is used in the three cases. Note that although the equations look similar to the ones discussed in Chapter 2, the nonlinear terms are of a different form. This is a crucial difference. In dimensionless form, these systems are governed by the following equations:

1. Primary system with external excitation:

$$x'' + \kappa_1 x' + x + \gamma(x^2 + y^2)x = a \cos \eta \tau,$$
$$y'' + \kappa_2 y' + q^2 y + \gamma(x^2 + y^2)y = 0. \qquad (3.1.1)$$

2. Primary system with parametric excitation:

$$x'' + \kappa_1 x' + (1 + a \cos 2\eta\tau)x + \gamma(x^2 + y^2)x = 0,$$
$$y'' + \kappa_2 y' + q^2 y + \gamma(x^2 + y^2)y = 0. \qquad (3.1.2)$$

3. Primary system with self-excitation:

$$x'' - (\beta - \delta x^2)x' + x + \gamma(x^2 + y^2)x = 0,$$
$$y'' + \kappa y' + q^2 y + \gamma(x^2 + y^2)y = 0. \qquad (3.1.3)$$

In all the preceding equations, κ_1, κ_2, γ, β, δ, κ, a, η, and q are positive constants. The first equation in each system represents the primary system, the second one the secondary system. In both equations of each system the nonlinear coupling term is of the third order. In all three systems, when we take $\kappa_1 = \kappa_2 = \kappa = a = \beta = \delta = 0$, each example becomes a Hamiltonian system with potential

$$V(x, y) = \tfrac{1}{2}(x^2 + y^2) + \tfrac{1}{4}(x^2 + y^2)^2.$$

Hamiltonian systems with such a symmetric potential arise quite often in applications, and our three examples can be seen as perturbations of such a Hamiltonian. Also, as mentioned earlier, each of these systems can represent a single-mass system with 2 degrees of freedom that is autoparametrically excited in the direction of the y axis.

The analysis of the semitrivial solution and its stability runs exactly along the same lines as in Chapters 1 and 2; we leave this to the reader for the system with parametric excitation. The discussions of the case of a system with an externally excited primary system is based on results obtained by Nabergoj and Tondl (1994). Another example of a self-excited system is analysed in Nabergoj and Tondl (1996).

3.2 A Primary System with External Excitation

Consider the system

$$\begin{aligned} x'' + \kappa_1 x' + x + \gamma(x^2 + y^2)x &= a\cos\eta\tau, \\ y'' + \kappa_2 y' + q^2 y + \gamma(x^2 + y^2)y &= 0, \end{aligned} \quad (3.2.1)$$

with positive coefficients κ_1, κ_2, γ, a, η, and q. We are particularly interested in the case of resonance, i.e., when η is close to 1. In that case we make the following scaling: $\kappa_1 = \varepsilon\hat{\kappa}_1$, $\kappa_2 = \varepsilon\hat{\kappa}_2$, $\gamma = \varepsilon\hat{\lambda}$, and $a = \varepsilon^2 \hat{a}$. From this point on, we will drop all the hats.

3.2.1 Semitrivial Solution and Stability

To study the semitrivial solution we have to solve

$$x'' + \varepsilon\kappa_1 x' + x + \varepsilon\gamma x^3 = \varepsilon^2 a \cos\eta\tau. \quad (3.2.2)$$

Assuming that η is near to 1, we propose an approximation of the solution of system (3.1.1) by

$$\begin{aligned} x_0(\tau) &= a_0 \cos\eta\tau + b_0 \sin\eta\tau, \\ y_0(\tau) &= 0. \end{aligned} \quad (3.2.3)$$

Elementary Discussion of Single-Mass Systems

The values of a_0 and b_0 can be determined with the harmonic balance method or averaging and results in the conditions

$$[1 - \eta^2 + \tfrac{3}{4}\gamma(a_0^2 + b_0^2)]a_0 + \kappa_1 \eta b_0 = a,$$
$$-\kappa_1 \eta a_0 + [1 - \eta^2 + \tfrac{3}{4}\gamma(a_0^2 + b_0^2)]b_0 = 0. \qquad (3.2.4)$$

When the amplitude of the semitrivial solution $R_0 = (a_0^2 + b_0^2)^{1/2}$ is introduced, Eqs. (3.2.4) can be transformed into

$$\eta_{1,2}^2 = 1 + \tfrac{3}{4}\gamma R_0^2 - \tfrac{1}{2}\kappa_1^2 \pm \left[\frac{a^2}{R_0^2} - (1 + \tfrac{3}{4}\gamma R_0^2)\kappa_1^2 + \tfrac{1}{4}\kappa_1^4\right]^{1/2} \qquad (3.2.5)$$

from which the behaviour of $\eta(R_0)$ and $R_0(\eta)$ can be determined.

For the determination of the stabilty of system (3.2.3), the perturbative solution

$$x = x_0 + u,$$
$$y = y_0 + v, \qquad (3.2.6)$$

is inserted into system (3.2.1). Thus, in the first approximation, the following variational equations are obtained:

$$u'' + \kappa_1 u' + u + \tfrac{3}{2}\gamma[a_0^2 + b_0^2 + (a_0^2 - b_0^2)\cos 2\eta\tau + 2a_0 b_0 \sin 2\eta\tau]u = 0,$$
$$v'' + \kappa_2 v' + q^2 v + \tfrac{1}{2}\gamma[a_0^2 + b_0^2 + (a_0^2 - b_0^2)\cos 2\eta\tau + 2a_0 b_0 \sin 2\eta\tau]v = 0.$$

These equations are independent of each other and are both of Mathieu type. Using the approximation of the solution that is valid on the boundary of the main instability domain,

$$u(\tau) = U_c \cos \eta\tau + U_s \sin \eta\tau,$$
$$v(\tau) = V_c \cos \eta\tau + V_s \sin \eta\tau,$$

we obtain the following equations:

$$\{1 - \eta^2 + \tfrac{3}{2}\gamma[a_0^2 + b_0^2 + \tfrac{1}{2}(a_0^2 - b_0^2)]\}U_c + (\kappa_1\eta + \tfrac{3}{2}\gamma a_0 b_0)U_s = 0,$$
$$(\tfrac{3}{2}\gamma a_0 b_0 - \kappa_1\eta)U_c + \{1 - \eta^2 + \tfrac{3}{2}\gamma[a_0^2 + b_0^2 - \tfrac{1}{2}(a_0^2 - b_0^2)]\}U_s = 0,$$
$$\{q^2 - \eta^2 + \tfrac{1}{2}\gamma[a_0^2 + b_0^2 + \tfrac{1}{2}(a_0^2 - b_0^2)]\}V_c + (\kappa_2\eta + \tfrac{1}{2}\gamma a_0 b_0)V_s = 0,$$
$$(\tfrac{1}{2}\gamma a_0 b_0 - \kappa_2\eta)V_c + \{q^2 - \eta^2 + \tfrac{1}{2}\gamma[a_0^2 + b_0^2 - \tfrac{1}{2}(a_0^2 - b_0^2)]\}V_s = 0.$$

From the condition for the nontrivial solution of U_c and U_s, we deduce the following relation:

$$\eta_{1,2}^2 = 1 + \tfrac{3}{2}\gamma R^2 - \tfrac{1}{2}\kappa_1^2 \pm \left[\tfrac{9}{16}\gamma^2 R^4 - \left(1 + \tfrac{3}{2}\gamma R^2\right)\kappa_1^2 + \tfrac{1}{4}\kappa_1^4\right]^{1/2}, \quad (3.2.7)$$

where $R = (a_0^2 + b_0^2)^{1/2}$ is the stability boundary value. This relation is identical to the relation obtained when the stability of the Duffing equation is investigated. The first equation of system (3.2.1) reduces to the Duffing equation for $y(\tau) = 0$. In the interval of the excitation frequency of the primary system, where three amplitudes exist, the highest and the lowest amplitudes correspond to stable solutions.

The analogous condition for the nontrivial solution of V_c and V_s yields the relation

$$\eta_{1,2}^2 = q^2 + \tfrac{1}{2}\gamma R^2 - \tfrac{1}{2}\kappa_2^2 \pm \left[\tfrac{1}{16}\gamma^2 R^4 - \left(q^2 + \tfrac{1}{2}\gamma R^2\right)\kappa_2^2 + \tfrac{1}{4}\kappa_2^4\right]^{1/2}. \quad (3.2.8)$$

Figure 3.1 shows both the frequency response curve $R_0(\eta)$ given by Eq. (3.2.5) and the stability boundary curves $R(\eta)$ for the following parameter values: $\kappa_1 = \kappa_2 = 0.10$, $a = 1.00$, $\gamma = 0.02$, and $q = 1.10$. In the figure, the stability boundary curve of Figure 3.1(a) has been obtained from Eq. (3.2.7), whereas the curve of Figure 3.1(b) was obtained from Eq. (3.2.8). We can see that there exists an interval of the excitation frequency η where the semitrivial solution is not stable and autoparametric resonance is initiated. To illustrate the influence of the tuning coefficient, the amplitude R_0 of the stable semitrivial solution, as a function of the frequency η and the coefficient q, is presented in Figure 3.2 (for $\kappa_1 = \kappa_2 = 0.10$, $a = 1.00$, and $\gamma = 0.02$).

3.2.2 Domains of Stability in Parameter Space

Domains of attraction for 2 degrees of freedom are not so easy to display. Following Schmidt and Tondl (1986) and Nabergoj and Tondl (1994), we proceed as follows. The differential equations of motion are solved numerically for initial conditions that lead to a given stable solution with no transient. At a certain time τ_0, a disturbance is applied during an interval of time T_0 and the subsequent motion after time $\tau_0 + T_0$ is

Elementary Discussion of Single-Mass Systems

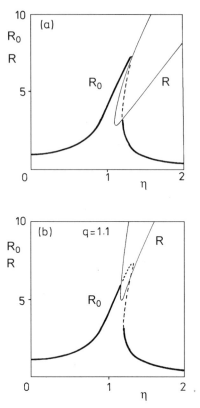

Figure 3.1: Vibration amplitude curve R_0 corresponding to the semitrivial solution (stable solution, heavy solid curves; unstable solution, dashed curves) and the stability boundary curve R (light solid curves) as a function of the excitation frequency η: (a) according to condition (3.2.7), (b) according to condition (3.2.8). The following values have been used: $\kappa_1 = \kappa_2 = 0.10$, $a = 1.00$, $\gamma = 0.02$, $q = 1.10$.

observed. The disturbance will be characterised by two parameters only: the amplitude of the disturbance and the length of its time interval. These parameters are used in a diagram to indicate to where the solutions are evolving. We take, for instance, a sinusoidal pulse,

$$P(\tau) = A \sin\left[\frac{\pi}{T_0}(\tau - \tau_0)\right], \qquad (3.2.9)$$

and use in our diagrams $T = 2\pi/\eta$ and $\theta = T_0/T$. We vary the disturbance stepwise by changing both the parameters A and T_0. White areas show the pulse characteristics that lead to the semitrivial solution, and

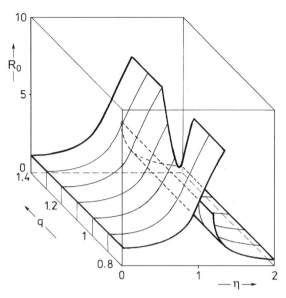

Figure 3.2: Axonometric representation of the stable vibration amplitude R_0 corresponding to the semitrivial solution as a function of the excitation frequency η and tuning ratio q. The following values have been used: $\kappa_1 = \kappa_2 = 0.10$, $a = 1.00$, $\gamma = 0.02$.

dotted areas indicate a return to a nontrivial solution. The diagram in Figure 3.3 refers to starting conditions at the semitrivial solution; the diagram in Figure 3.4 refers to starting conditions at the nontrivial solution. Note that this is a specific way of studying domains of attraction; more research is needed to explore such methods.

3.3 A Primary System with Self-Excitation

We consider the following system in dimensionless form:

$$x'' - (\beta - \delta x^2)x' + x + \gamma(x^2 + y^2)x = 0,$$
$$y'' + \kappa y' + q^2 y + \gamma(x^2 + y^2)y = 0, \quad (3.3.1)$$

in which β, δ, γ, and κ are small, positive parameters and $q \approx 1$. In contrast to system (2.1.3), we have a van der Pol-type self-excitation instead of a Rayleigh type. Also the nonlinearities and resonances are different. As we shall be more explicit here in our calculations, we

Elementary Discussion of Single-Mass Systems

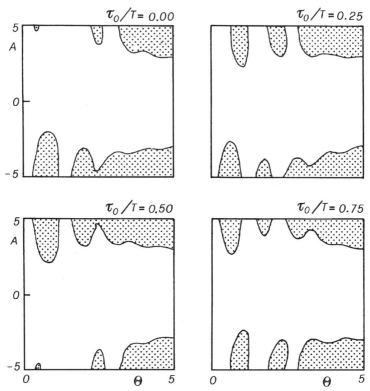

Figure 3.3: Domains of attraction when the starting conditions correspond to the semitrivial solution. Dotted areas represent conditions that lead to the nontrivial solution.

replace β, δ, γ, and κ with $\varepsilon\beta$, $\varepsilon\delta$, $\varepsilon\gamma$, and $\varepsilon\kappa$; we also put $q^2 - 1 = \varepsilon\sigma$. Our system becomes

$$x'' - \varepsilon(\beta - \delta x^2)x' + x + \varepsilon\gamma(x^2 + y^2)x = 0,$$
$$y'' + \varepsilon\kappa y' + (1 + \varepsilon\sigma)y + \varepsilon\gamma(x^2 + y^2)y = 0. \quad (3.3.2)$$

The semitrivial solution satisfies

$$x'' - \varepsilon(\beta - \delta x^2)x' + x + \varepsilon\gamma x^3 = 0, \quad (3.3.3)$$

with $y(\tau) = y_0(\tau) = 0$. As in Section 9.2, here we introduce phase-amplitude variables $x_0(\tau) = R_0 \cos(\tau + \phi)$, $x_0'(\tau) = -R_0 \sin(\tau + \phi)$.

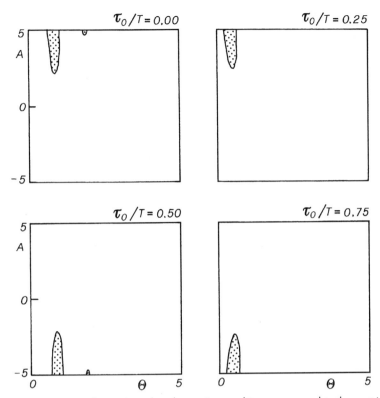

Figure 3.4: Domains of attraction when the starting conditions correspond to the nontrivial solution. Dotted areas represent conditions that lead to the semitrivial solution.

For the averaged equation from (3.3.3) we find

$$R_0' = \tfrac{1}{2}\varepsilon R_0 \left(\beta - \tfrac{1}{4}\delta R_0^2\right),$$
$$\phi' = \tfrac{3}{8}\varepsilon \gamma R_0^2, \qquad (3.3.4)$$

which produces for the periodic solution the condition

$$\beta = \tfrac{1}{4}\delta R_0^2. \qquad (3.3.5)$$

Solving $\phi(\tau)$ from system (3.3.4) and using Eq. (3.3.5), we find as an $\mathcal{O}(\varepsilon)$ approximation, valid on the time scale $1/\varepsilon$,

$$x_0(\tau) = 2 \left(\frac{\beta}{\delta}\right)^{1/2} \cos\left(1 + \frac{3}{2}\varepsilon\gamma\frac{\beta}{\delta}\right)\tau. \qquad (3.3.6)$$

The harmonic balance method yields the same expression.

Elementary Discussion of Single-Mass Systems

To study the stability of the semitrivial solution in full system (3.3.2), we consider the perturbed solutions

$$x = x_0(\tau) + u, \quad y = y_0(\tau) + v. \tag{3.3.7}$$

This produces, on linearisation,

$$u'' - \varepsilon[\beta - \delta x_0^2(\tau)]u' + [1 + 2\varepsilon\delta x_0(\tau)x_0'(\tau) + 3\varepsilon\gamma x_0^2(\tau)]u = 0,$$
$$v'' + \varepsilon\kappa v' + [1 + \varepsilon\sigma + \varepsilon\gamma x_0^2(\tau)]v = 0. \tag{3.3.8}$$

These equations are of Mathieu type and can be treated as discussed in Section 9.5. If we put $s = (1 + \frac{3}{2}\varepsilon\gamma\frac{\beta}{\delta})\tau$, the last equation of system (3.3.8) becomes, to $\mathcal{O}(\varepsilon)$,

$$\frac{d^2v}{ds^2} + v = -\varepsilon\kappa\frac{dv}{ds} + \varepsilon\left(\gamma\frac{\beta}{\delta} - \sigma\right)v - 2\varepsilon\gamma\frac{\beta}{\delta}\cos 2s\; v. \tag{3.3.9}$$

By applying the theory of Section 9.5, we find the stability boundary:

$$q^2 = 1 + \varepsilon\gamma\frac{\beta}{\delta} \pm \varepsilon\sqrt{\gamma^2\frac{\beta^2}{\delta^2} - \kappa^2}. \tag{3.3.10}$$

A somewhat longer calculation for the first equation of system (3.3.8) yields that in this case no instability domain exists. It is instructive to see that we can obtain this result in a much simpler way by averaging, as this also enables us to study the nontrivial solutions in the case of autoparametric resonance. We show this in some detail in Subsection 3.3.1.

3.3.1 Nontrivial Solutions in the Case of a Self-Excited Primary System

To study system (3.3.2), we introduce amplitude-phase coordinates:

$$x = R_1\cos(\tau + \phi_1), \quad y = R_2\cos(\tau + \phi_2),$$
$$x' = -R_1\sin(\tau + \phi_1), \quad y' = -R_2\sin(\tau + \phi_2). \tag{3.3.11}$$

Averaging the resulting equations for R_1, R_2, ϕ_1, and ϕ_2, we find

$$R_1' = \varepsilon\left[\tfrac{1}{2}R_1(\beta - \tfrac{1}{4}\delta R_1^2) + \tfrac{1}{8}\gamma R_1 R_2^2 \sin 2(\phi_1 - \phi_2)\right],$$
$$\phi_1' = \varepsilon\left[\tfrac{3}{8}\gamma R_1^2 + \tfrac{1}{4}\gamma R_2^2 + \tfrac{1}{8}\gamma R_2^2 \cos 2(\phi_1 - \phi_2)\right],$$
$$R_2' = \varepsilon\left[-\tfrac{1}{2}\kappa R_2 - \tfrac{1}{8}\gamma R_1^2 R_2 \sin 2(\phi_1 - \phi_2)\right],$$
$$\phi_2' = \varepsilon\left[\tfrac{1}{2}\sigma + \tfrac{3}{8}\gamma R_2^2 + \tfrac{1}{4}\gamma R_1^2 + \tfrac{1}{8}\gamma R_1^2 \cos 2(\phi_1 - \phi_2)\right]. \tag{3.3.12}$$

At this point we note that, on putting $R_2 = 0$ in system (3.3.12), we recover the semitrivial solution. It is also immediately clear from the first equation of system (3.3.12) that, on considering the stability of the semitrivial solution in the full system, we have no instability in the (R_1, ϕ_1) dimension of motion of the single mass.

If we put $\chi = \phi_1 - \phi_2$ for the combination angle, we reduce system (3.3.12) to three equations, i.e., the equations for R_1 and R_2 and

$$\chi' = -\tfrac{1}{2}\varepsilon\sigma + \tfrac{1}{8}\varepsilon\gamma(R_1^2 - R_2^2) - \tfrac{1}{8}\varepsilon\gamma(R_1^2 - R_2^2)\cos 2\chi. \qquad (3.3.13)$$

The simplest nontrivial periodic solution arises if

$$R_1' = R_2' = 0, \quad R_1 > 0, \quad R_2 > 0. \qquad (3.3.14)$$

This leads to

$$(\beta - \tfrac{1}{4}\delta R_1^2) + \tfrac{1}{4}\gamma R_2^2 \sin 2\chi = 0,$$
$$\kappa + \tfrac{1}{4}\gamma R_1^2 \sin 2\chi = 0. \qquad (3.3.15)$$

Eliminating $\sin 2\chi$ from system (3.3.15), we find the relation

$$R_1^2(\beta - \tfrac{1}{4}\delta R_1^2) = \kappa R_2^2, \quad R_1, R_2 > 0. \qquad (3.3.16)$$

Condition (3.3.16) shows that we have

$$0 < R_1 < 2\sqrt{\beta/\delta}. \qquad (3.3.17)$$

In general, R_1 will be smaller than the amplitude of the semitrivial solution; see Figure 6.7 in Chapter 6. Moreover, we note that we may have a nontrivial solution if $0 < \kappa < \beta$. If $\kappa > \beta$, condition (3.3.16) cannot be satisfied. If R_1 and R_2 are constant, χ has to be constant as well, so the remaining condition to be considered is $\chi' = 0$, or

$$\tfrac{1}{4}\gamma(R_1^2 - R_2^2)\cos 2\chi = \sigma + \tfrac{1}{4}\gamma(R_1^2 - R_2^2). \qquad (3.3.18)$$

We can eliminate R_2^2 from Eq. (3.3.18) by using condition (3.3.16). Also, the second equation of system (3.3.15) yields

$$\sin 2\chi = -\frac{4\kappa}{\gamma R_1^2}. \qquad (3.3.19)$$

The elimination of χ from Eqs. (3.3.18) and (3.3.19) gives us a polynomial equation in R_1^2 of degree 6. We do not analyse this polynomial equation in general but we make a number of observations:

Elementary Discussion of Single-Mass Systems

1. Equation (3.3.19) contains a condition on R_1^2. We have clearly

$$R_1^2 \geq \frac{4\kappa}{\gamma} \qquad (3.3.20)$$

   for nontrivial solutions to exist.

2. The nontrivial periodic solutions are phase locked, $\phi_1(t) - \phi_2(t) =$ const. If they exist, there are in general two of them, differing by a phase of π.

3. Near the parameter value $\kappa = \beta$, interesting bifurcational behaviour can be observed. If β is fixed and the value of κ passes through β from above, we have a pitchfork bifurcation. At $\kappa = \beta$ the curves in Figure 6.7 are tangent at $R_1 = R_2 = 0$. Decreasing κ a little produces nontrivial solutions with R_1 and R_2 far way from zero.

4. Note that the conditions that were obtained for the existence of nontrivial solutions are different from condition (3.3.10) for the stability–instability of the semitrivial solution. For instance, it follows from condition (3.3.10) that $q^2 \leq 1$ implies stability of the semitrivial solution. As we shall see, nontrivial solutions may exist for these parameter values, showing that autoparametric resonance can be generated outside the instability domain of the semitrivial solution.

5. No damping, $\kappa = 0$. It is interesting to look at the limit case $\kappa = 0$. Then system (3.3.15) and Eq. (3.3.13) produce $\sin 2\chi = 0$ and $\cos 2\chi = -1$, so $\chi = \frac{1}{2}\pi$, $\chi = \frac{3}{2}\pi$. No energy is absorbed from the self-excited primary system by damping of the secondary system, and we have $R_1 = 2\sqrt{\beta/\delta}$.

   As $\cos 2\chi = \pm 1$, Eq. (3.3.18) yields $\frac{1}{2}\gamma(R_1^2 - R_2^2) = \sigma$ or $R_2^2 = 4(\beta/\delta) - 2(\sigma/\gamma)$. Note that the right-hand side has to be positive. Also, $R_2 = R_1$ at exact internal resonance ($\sigma = 0$), and R_2 decreases with the detuning σ. The linearisation of the right-hand sides of Eqs. (3.3.12) at the periodic solution has a cubic characteristic equation with positive coefficients. So we have stability of the periodic solution.

6. Exact internal resonance, $\sigma = 0$. In this case the semitrivial solution is stable; see Eq. (3.3.10). From Eq. (3.3.13) we find that

$$(R_1^2 - R_2^2)(1 - \cos 2\chi) = 0. \qquad (3.3.21)$$

   If $\cos 2\chi = 1$, then $\sin 2\chi = 0$ and system (3.3.12) leads to $R_2 = 0$

(unless $\kappa = 0$; see observation 5). We discard this case and are left with $R_1 = R_2$. It follows from Eqs. (3.3.16) and (3.3.15) that

$$R_1 = R_2 = 2\sqrt{\frac{\beta - \kappa}{\delta}}, \quad \sin 2\chi = \frac{\delta \kappa}{\gamma(\kappa - \beta)}. \qquad (3.3.22)$$

Note that again we have $0 < \beta < \kappa$ and, moreover, the condition

$$\frac{\delta \kappa}{\gamma(\kappa - \beta)} < 1.$$

Although in the above-mentioned cases it is (in principle) not difficult to establish the stability of the solution, the calculations tend to become long. If necessary, the reader can perform these calculations.

In Chapter 2 we studied an autoparametric system with a self-excited primary system of the Rayleigh type. The elementary results with regards to semitrivial and nontrivial solutions are similar. This chapter shows that we are able to give a more detailed analysis of the nontrivial solutions by approximation methods. We extend this for systems with a self-excited primary system in Chapter 7.

Chapter 4

Mass–Spring–Pendulum Systems

In this chapter we give a detailed account of a system that has been widely studied in the literature. We consider the semitrivial solutions and their stability and various nontrivial solutions including quenching, and we observe various bifurcations. The nontrivial solutions that arise for certain excitation frequencies can undergo a Hopf bifurcation to a limit-cycle motion. These limit cycles in their turn can lead to period doubling followed by chaos.

4.1 The Resonantly Driven Mass–Spring–Pendulum System

The mechanical system treated in this section has 2 degrees of freedom and is schematically represented in Figure 4.1. It consists of a mass M mounted on a linear spring, which is carrying a simple pendulum. The pendulum has mass m, attached to a hinged massless rod having length l. In the figure y denotes the coordinate of the mass M and φ is the angular deflection of the pendulum. Moreover, the mass M is excited by a harmonic force $P \cos \omega t$. In certain applications the forcing P may depend on the frequency ω. We also assume linear viscous damping.

There are three frequencies associated with this system: the natural frequencies of the pendulum and of the linear mass–spring system and

Mass–Spring–Pendulum Systems

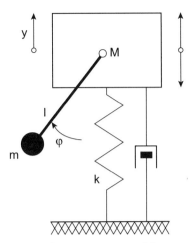

Figure 4.1: The mass–spring–pendulum system.

the frequency of the driving force. These frequencies are denoted by ω_0, ω_1, and ω, respectively.

The most interesting situation and the one that has been most widely studied in the literature occurs when the natural frequencies of the mass–spring system (primary system) and the pendulum (secondary system) are in 2:1 resonance and the frequency of the external forcing is in 1:1 resonance with the natural frequency of the primary system. The resonant motion of the primary system then acts as a parametric excitation on the motion of the pendulum. Conversely, the motion of the pendulum influences the motion of the mass–spring system. It is possible that energy from the mass–spring motion is transferred to the motion of the pendulum. When this happens, the pendulum effectively acts as a vibration absorber. In many engineering applications, a system can be modelled by a linear mass–spring system, subjected to a periodic forcing. When this forcing is resonant, a large-amplitude response of the elastically mounted mass occurs, which might be undesirable. In this chapter it is shown that, by the addition of a suitably tuned subsystem (in this case a pendulum), the amplitude of the mass–spring motion can be greatly reduced.

Previous studies of this autoparametric system include Haxton and Barr (1972), Hatwal et al. (1983), Bajaj et al. (1994), Ruijgrok (1995), and Banerjee et al. (1996). In this chapter we give a summary of the results of these authors and we present some new results.

45

4.1.1 Equations of Motion

The equations of motion are derived in the following way. The kinetic energy of this system is given by

$$T = \frac{1}{2}M\dot{y}^2 + \frac{1}{2}m(\dot{y} + l\dot{\varphi}\sin\varphi)^2 + \frac{1}{2}m(l\dot{\varphi}\cos\varphi)^2, \quad (4.1.1)$$

and the potential energy is given by

$$U = mgl(1 - \cos\varphi) + \frac{1}{2}ky^2. \quad (4.1.2)$$

From the Lagrange equations we find

$$(M+m)\ddot{y} + b\dot{y} + ky + ml(\ddot{\varphi}\sin\varphi + \dot{\varphi}^2\cos\varphi) = P\cos\omega t,$$
$$ml^2\ddot{\varphi} + c\dot{\varphi} + mgl\sin\varphi + ml\ddot{y}\sin\varphi = 0, \quad (4.1.3)$$

where b and c are the damping coefficients and g is the acceleration of gravity. Equations (4.1.3) can be written in dimensionless form as

$$x'' + \kappa x' + \frac{\omega_1^2}{\omega^2}x + \mu(\varphi''\sin\varphi + \varphi'^2\cos\varphi) = a\cos\tau,$$
$$\varphi'' + \kappa_0\varphi' + \frac{\omega_0^2}{\omega^2}\sin\varphi + x''\sin\varphi = 0, \quad (4.1.4)$$

where the following transformations have been used: $x = y/l$, $\omega t = \tau$, $\kappa = b/\omega(M+m)$, $\kappa_0 = c/\omega ml^2$, $\omega_1^2 = k/(M+m)$, $\omega_0^2 = g/l$, $\mu = m/(M+m)$, and $a = P/l\omega^2(M+m)$. The prime indicates the derivative with respect to the dimensionless time variable τ.

As was mentioned, we are particulary interested in the case in which $\omega_1 \approx \omega$ and $\omega_0 \approx \frac{1}{2}\omega$.

4.1.2 Various Classes of Solutions

A 2 degrees of freedom system with forcing, such as system (4.1.4), is in general very rich in types of solutions; there exist stationary solutions, periodic solutions, invariant tori, and chaos. Here we identify some of the more simple solutions of system (4.1.4).

First we have the semitrivial solution, defined by $\varphi = \varphi' = 0$, and (x, x'), the 2π-periodic solution of

$$x'' + \kappa x' + \frac{\omega_1^2}{\omega^2}x = a\cos\tau. \quad (4.1.5)$$

We use asymptotic methods to calculate the stability of this solution; therefore we require that this periodic solution be small in a sense to be specified. The (positive) damping coefficient κ is usually considered small, and we use this coefficient to define the small parameter of the system by scaling the damping coefficients $\kappa = \varepsilon \kappa_1$, with $0 < \varepsilon \ll 1$ and $\kappa_1 > 0$.

As was mentioned, we assume that the primary system is resonantly excited, i.e., that $\omega_1 \approx \omega$. Writing $\omega_1^2/\omega^2 = 1 + \varepsilon \sigma_1$, where σ_1 determines the detuning from exact resonance, we obtain the periodic solution of Eq. (4.1.5), which is given by

$$x(t) = R_0 \cos(\tau + \psi_0), \qquad (4.1.6)$$

where the amplitude R_0 and the phase ψ_0 are given by

$$R_0 = \frac{a}{\varepsilon\left(\sigma_1^2 + \kappa_1^2\right)^{1/2}}, \quad \psi_0 = \arctan\left(\frac{\kappa_1}{\sigma_1}\right). \qquad (4.1.7)$$

It follows that we must take $a = \mathcal{O}(\varepsilon^2)$ to guarantee that the size of R_0 is of $\mathcal{O}(\varepsilon)$. In the following paragraphs we study the stability of the semitrivial solutions and the nontrivial solutions that bifurcate from them.

When we take $a = \mathcal{O}(\varepsilon)$, the amplitude of the periodic solution of Eq. (4.1.5) has grown to $\mathcal{O}(1)$. However, we show that a periodic solution of autoparametric system (4.1.4) exists such that $x = \mathcal{O}(\varepsilon)$ and $\varphi = \mathcal{O}(\sqrt{\varepsilon})$. This is a highly remarkable solution. The mass–spring system is driven resonantly, but it yields a response of (approximately) the same amplitude as that of the driving force. This is of course possible only because we added a pendulum to the system, with an appropriate eigenfrequency. This pendulum absorbs much of the energy of the resonantly forced primary system, accounting for its relatively large motion of $\mathcal{O}(\sqrt{\varepsilon})$. We call this solution the strongly quenched solution, as the response of the mass–spring system is made small, i.e., is quenched, when the pendulum subsystem is added. In paragraphs 4.2 and 4.3 we consider the stability and the bifurcations of this solution.

Finally, there are many other solutions of Eqs. (4.1.4). In particular, there can exist rotating periodic solutions, where $\varphi'(\tau) > 0$ for all time τ. These solutions are discussed at the end of this chapter.

4.1.3 Stability of the Semitrivial Solution

Using the small parameter ε, we scale $\kappa = \varepsilon \kappa_1$, $\kappa_0 = \varepsilon \kappa_2$, and $a = \varepsilon^2 \hat{a}$. To study the stability of the semitrivial solution, we rewrite Eqs. (4.1.4) and scale $x = \varepsilon \hat{x}$ and $\varphi = \varepsilon \hat{\varphi}$. Dropping the hats and introducing a second detuning by $\omega_0^2/\omega^2 = \frac{1}{4} + \frac{1}{2}\varepsilon\sigma_2$ yields

$$x'' + x + \varepsilon\left(\kappa_1 x' + \sigma_1 x - \tfrac{1}{4}\mu\varphi^2 + \mu\varphi'^2\right) = \varepsilon a \cos\tau + \mathcal{O}(\varepsilon^2),$$
$$\varphi'' + \tfrac{1}{4}\varphi + \varepsilon\left(\kappa_2 \varphi' + \tfrac{1}{2}\sigma_2\varphi - x\varphi\right) = 0 + \mathcal{O}(\varepsilon^2). \quad (4.1.8)$$

Putting $x(t) = R_0 \cos(\tau + \psi_0) + u$ and $\varphi(t) = 0 + v(t)$ and linearising around the semitrivial solution then yields the variational equations

$$u'' + u + \varepsilon(\kappa_1 u' + \sigma_1 u) = 0,$$
$$v'' + \tfrac{1}{4}v + \varepsilon\left[\kappa_2 v' + \tfrac{1}{2}\sigma_2 v - r_0 \cos(\tau + \psi_0)v\right] = 0. \quad (4.1.9)$$

The first equation of Eqs. (4.1.9) shows that $u \to 0$. Therefore the stability is solely determined by the second equation of Eqs. (4.1.9). This is a Mathieu equation, and by using the same method as in Chapter 2 (see also Chapter 9) to determine the boundaries of the stability domains, we find that the semitrivial solution is stable if

$$R_0^2 \leq \kappa_2^2 + \sigma_2^2. \quad (4.1.10)$$

Using the expression for R_0 from Eqs. (4.1.7), we find the following condition for the stability of the semitrivial solution:

$$a^2 \leq \left(\kappa_1^2 + \sigma_1^2\right)\left(\kappa_2^2 + \sigma_2^2\right). \quad (4.1.11)$$

We note that the smaller the pendulum damping, the larger the instability domain of the semitrivial solution.

4.2 Nontrivial Solutions

We now look for nontrivial periodic solutions of system (4.1.8) near $\varphi = 0$. Introducing the transformations

$$x = R_1 \cos(\tau + \psi_1), \quad \varphi = R_2 \cos(\tfrac{1}{2}\tau + \psi_2),$$
$$\dot{x} = -R_1 \sin(\tau + \psi_1), \quad \dot{\varphi} = -\tfrac{1}{2}R_2 \sin(\tfrac{1}{2}\tau + \psi_2),$$

yields equations for R_1, R_2 and ψ_1, ψ_2 that are in the standard form for averaging. After averaging, these equations become

$$\dot{R}_1 = \tfrac{1}{2}\varepsilon\left[-\kappa_1 R_1 - \tfrac{1}{4}\mu R_2^2 \sin(\psi_1 - 2\psi_2) - a\sin\psi_1\right],$$
$$R_1\dot{\psi}_1 = \tfrac{1}{2}\varepsilon\left[\sigma_1 R_1 - \tfrac{1}{4}\mu R_2^2 \cos(\psi_1 - 2\psi_2) - a\cos\psi_1\right],$$
$$\dot{R}_2 = \tfrac{1}{2}\varepsilon[-\kappa_2 R_2 + R_1 R_2 \sin(\psi_1 - 2\psi_2)],$$
$$R_2\dot{\psi}_2 = \tfrac{1}{2}\varepsilon[\sigma_2 R_2 - R_1 R_2 \cos(\psi_1 - 2\psi_2)]. \tag{4.2.1}$$

Equating the right-hand sides of Eqs. (4.2.1) to zero yields the fixed points of these equations, which correspond with 4π-periodic solutions of the original equations (4.1.8). The last two equations of Eqs. (4.2.1) yield a simple expression for the amplitude R_1:

$$R_1 = \left(\sigma_2^2 + \kappa_2^2\right)^{1/2}. \tag{4.2.2}$$

A slightly longer calculation eventually yields a quadratic equation for R_2^2:

$$r^2 + 2(\kappa_1\kappa_2 - \sigma_1\sigma_2)r + \left(\kappa_1^2 + \sigma_1^2\right)\left(\kappa_2^2 + \sigma_2^2\right) - a^2 = 0, \tag{4.2.3}$$

where we have abbreviated $r = \tfrac{1}{4}\mu R_2^2$.

It is not difficult to show that Eq. (4.2.3) has two positive solutions if

$$(\kappa_1\sigma_2 + \kappa_2\sigma_1)^2 < a^2 < \left(\kappa_1^2 + \sigma_1^2\right)\left(\kappa_2^2 + \sigma_2^2\right),$$
$$\kappa_1\kappa_2 - \sigma_1\sigma_2 > 0,$$

one positive solution if,

$$a^2 > \left(\kappa_1^2 + \sigma_1^2\right)\left(\kappa_2^2 + \sigma_2^2\right),$$

and no positive solutions otherwise.

Fixing κ_1, κ_2, and $\sigma_2 > 0$, we can make a diagram in (σ_1, a) space; see Figure 4.2. Note that it follows from condition (4.1.11) that the semitrivial solution is unstable within region A and stable outside of it.

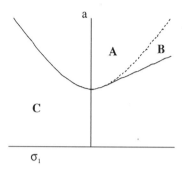

Figure 4.2: Existence of nontrivial solutions for fixed values of κ_1, κ_2, and $\sigma_2 > 0$. In region A there exists one nontrivial solution, in region B two, and in region C none. For $\sigma_2 < 0$ the diagram is reflected in the a axis.

4.2.1 Quenching and Its Relation to Pendulum Damping

At this point we compare the amplitude of the mass–spring motion for the semitrivial solution,

$$R_1^2 = \frac{a^2}{\sigma_1^2 + \kappa_1^2}, \qquad (4.2.4)$$

with the corresponding amplitude for the nontrivial solution,

$$R_1^2 = \sigma_2^2 + \kappa_2^2. \qquad (4.2.5)$$

An important observation that can be made from Eq. (4.2.5) is that in the case of the nontrivial solution, the amplitude of the mass–spring motion is independent of a (the amplitude of the external forcing). Increasing the amplitude of the forcing will result in the increase of the amplitude of the pendulum, but not in the increase of the amplitude of the spring motion. This phenomenon is usually referred to as quenching. The autoparametric pendulum in an oscillatory state can be used succesfully for the quenching of externally excited vibrations when the system is in resonance or close to it. This is in contrast to the use of a tuned absorber, which is often adopted in practice for quenching purposes. There the vibration can, at optimal tuning, also be diminished in the case in which the basic system is not in resonance.

Another result that follows from Eq. (4.2.5) is that the efficiency of the quenching depends on only the properties of the pendulum. In particular it can be seen that the smaller the pendulum damping, the smaller the amplitude of the mass–spring motion, i.e., the higher the efficiency of the quenching. This is similar to the case of using a tuned absorber.

4.2.2 Bifurcations and Stability of the Nontrivial Solutions

The results in this section can be found in more detail in Banerjee et al. (1996), Win-Min et al. (1994), and Ruijgrok (1995); in the first reference, attention is also given to the influence of increasing ε. The subsequent sections contain new results.

On the boundaries between the various regions in Figure 4.2 certain bifurcations occur, which we summarise below:

- On the boundary between A and B there occurs a subcritical period-doubling bifurcation of the (2π-periodic) semitrivial solution.
- On the boundary between A and C the semitrivial solution undergoes a supercritical period doubling.
- On the boundary between B and C there occurs a saddle-node bifurcation of fixed points in system (4.2.1). For original equations (4.1.8) this corresponds to a saddle-node bifurcation of 4π-periodic solutions.

Establishing the stability of the nontrivial solutions is not difficult, but the calculations can become lengthy. They lead to a fourth-degree eigenvalue equation, to which the Routh–Hurwitz condition can be applied. It follows that the stable and the unstable solutions, which are generated by the saddle-node bifurcation, remain stable and unstable, respectively, in the whole of region B. However, within region A the unique nontrivial solution can lose stability in a Hopf bifurcation. A more complete bifurcation diagram is given in Figure 4.3.

In region D the nontrivial solution is unstable. The solid line indicates a supercritical Hopf bifurcation to a stable two-frequency quasi-periodic solution, corresponding to a stable periodic solution of the averaged equations (4.2.1). The dotted curve indicates a subcritical Hopf bifurcation.

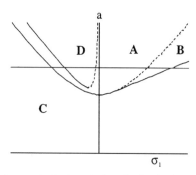

Figure 4.3: Bifurcation diagram in the (a, σ_1) plane for fixed values of κ_1, κ_2, and $\sigma_2 > 0$. The horizontal line indicates the fixed value a^* for which the response will be further investigated.

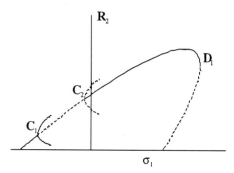

Figure 4.4: Response of R_2 against σ_1 for the fixed value of a^* indicated in Figure 4.3.

In Figure 4.4 we show the response of R_2 against σ_1 for the fixed value a^* indicated in Figure 4.3. When σ_1 crosses the boundary between regions C and A (see Figure 4.3), the stable response changes from the 2π-periodic semitrivial solution to a 4π-periodic nontrivial solution. This solution (see Figure 4.4) loses stability at point C_1, where it undergoes a supercritical Hopf bifurcation to a quasi-periodic solution. The dashed curve between the points C_1 and C_2 indicates that the nontrivial solution is not stable for the corresponding values of σ_1.

We remark that in Banerjee et al. (1996) it is shown numerically that for values of σ_1 between the Hopf bifurcation points C_1 and C_2 repeated period doublings can occur, leading to larger and larger periods. This suggests the presence of a homoclinic orbit, and we associate with it a Šilnikov bifurcation. This implies chaotic behaviour, which has been confirmed by numerical experiments. At point D_1, where σ_1 crosses from region B back to C, a hysteresis jump can occur.

Finally we remark that a diagram of the response of R_1 against σ_1 is less informative, as it follows from Eq. (4.2.5) that the R_1 components of the nontrivial solutions are independent of σ_1.

4.3 A Strongly Quenched Solution

In this section we consider the situation in which the amplitude of the excitation is of $\mathcal{O}(\varepsilon)$, in contrast to the preceding sections in which we assumed an amplitude of $\mathcal{O}(\varepsilon^2)$. In this case, it is easy to see that the amplitude of the semitrivial solution is of $\mathcal{O}(1)$. The semitrivial solution is in this case generally unstable, although we cannot show this by using asymptotic methods and we have to use numerical calculations.

There exists, however, a nontrivial solution for which the x component is of $\mathcal{O}(\varepsilon)$. This is remarkable, as it is the x component (which describes the motion of the mass–spring system) that is resonantly forced. In the absence of the pendulum subsystem the response of the mass–spring system would be of $\mathcal{O}(1)$. This solution is essentially the same solution that we called quenched in Section 4.2. It was noted there that the amplitude of this solution was independent of the amplitude of the external forcing. In this section we show that this solution survives even when the amplitude of the forcing is increased by an order of magnitude [from $\mathcal{O}(\varepsilon^2)$ to $\mathcal{O}(\varepsilon)$]. Because there is now an even more dramatic difference between the amplitudes of the semitrivial solution (large) and the nontrivial solution (small), we refer to the latter as a strongly quenched solution.

4.3.1 The Rescaled Equations

To investigate the above-mentioned solution, we rescale $a = \varepsilon \hat{a}$, $x = \sqrt{\varepsilon} \hat{x}$, and $\phi = \sqrt{\varepsilon} \hat{\phi}$, and expand Eqs. (4.1.4) up to $\mathcal{O}(\sqrt{\varepsilon})$ terms. After some rewriting, this yields (with the hats dropped)

$$x'' + x + \sqrt{\varepsilon}\left(-\tfrac{1}{4}\mu\varphi^2 + \mu\varphi'^2 - a\cos\tau\right) = \mathcal{O}(\varepsilon),$$
$$\varphi'' + \tfrac{1}{4}\varphi - \sqrt{\varepsilon}x\varphi = \mathcal{O}(\varepsilon). \qquad (4.3.1)$$

We observe that when we ignore the $\mathcal{O}(\varepsilon)$ terms on the right-hand side, these equations have an exact solution:

$$x = x_0 = 0, \quad \varphi = \varphi_0 = 2\sqrt{\frac{a}{\mu}} \sin \tfrac{1}{2}\tau. \qquad (4.3.2)$$

Equations (4.3.1) can be written as a four-dimensional system of ordinary differential equations, which we abbreviate as

$$z' = Lz + \sqrt{\varepsilon}F(z, \tau) + \varepsilon G(z, \tau). \qquad (4.3.3)$$

The vector $z \in \mathcal{R}^4$ has the usual interpretation: $(z_1, z_2, z_3, z_4) = (x, \dot{x}, \varphi, \dot{\varphi})$, and L is the linear part of the equation. L has eigenvalues $\pm i$ and $\pm \tfrac{1}{2}i$. The functions $F(z, \tau)$ and $G(z, \tau)$ contain time-dependent and nonlinear terms and are 2π periodic in τ. Solutions (4.3.2) are exact, simply harmonic 4π-periodic solutions of Eqs. (4.3.1) when the $\mathcal{O}(\varepsilon)$ terms are ignored. Therefore this solution corresponds to a 4π-periodic solution $z_0(\tau)$ of Eq. (4.3.3) such that $z_0' = Lz_0$ and $F[z_0(\tau), \tau] = 0$ for all $t \in \mathcal{R}$.

To analyse Eq. (4.3.3), we can put it in the standard form for averaging by transforming $z = e^{L\tau}w$. This leads to an equation for w:

$$w' = \sqrt{\varepsilon} e^{-L\tau} F(e^{L\tau}w, \tau) + \varepsilon e^{-L\tau} G(e^{L\tau}w, \tau)$$
$$:= \sqrt{\varepsilon} f(w, \tau) + \varepsilon g(w, \tau). \qquad (4.3.4)$$

Note that, because of the multiplication by the 4π-periodic matrix $e^{-L\tau}$, the functions $f(w, \tau)$ and $g(w, \tau)$ are 4π periodic in τ.

Define $w_0 = e^{-L\tau}z_0(\tau)$. Then, because z_0 is a solution of $z' = Lz$, w_0 is a constant, independent of τ. Also, because $F[z_0(\tau), \tau] = 0$ for all $t \in \mathcal{R}$, we have that $f(w_0, \tau) = 0$ for all $\tau \in \mathcal{R}$. Expanding $f(w, \tau)$ in a Fourier series, $f(w, \tau) = f_0(w) + \sum_{k \neq 0} f_k(w) e^{ik\tau}$, we note that, in particular, $f_0(w_0) = 0$.

Equation (4.3.4) is in the standard form for averaging. The truncated averaged equation has the form

$$w' = \sqrt{\varepsilon} f_0(w). \qquad (4.3.5)$$

A fixed point of this equation is given by $w = w_0$. From the theory of averaging [see Verhulst (1996)], it follows that Eq. (4.3.4) has a 4π-periodic solution $w(\tau) = w_0 + \mathcal{O}(\sqrt{\varepsilon})$, or equivalently, system (4.3.1) has a 4π-periodic solution $\hat{x}(\tau) = x_0 + \mathcal{O}(\sqrt{\varepsilon})$, $\varphi(\tau) = \varphi_0(\tau) + \mathcal{O}(\sqrt{\varepsilon})$, provided that the linear operator $D_w f(w_0)$ has no zero eigenvalues. Because $x_0 = 0$ and because we had already scaled the original x coordinate by $\sqrt{\varepsilon}$, this means that we have constructed a 4π-periodic solution $[x(t), \varphi(\tau)]$ of the original equations (4.1.4) such that $x(\tau) = \mathcal{O}(\varepsilon)$, which is the same order of magnitude as the amplitude of the excitation.

4.3.2 A Quasi-Degenerate Hopf Bifurcation

To complete the proof of the assertion that a strongly quenched solution exists, we need to calculate the eigenvalues of $D_w f(w_0)$. This will also provide information on the stability of the solution. Rather than performing the transformations described above to calculate $D_w f(w_0)$, we observe that we arrive at the same linear operator by writing $x = x_0 + u$, $\varphi = \varphi_0 + v$, linearising the equations for u and v, and then averaging the resulting equations. A short calculation shows that

$$D_w f(w_0) = \tfrac{1}{2} \sqrt{\frac{a}{\mu}} \begin{pmatrix} 0 & 0 & -\mu & 0 \\ 0 & 0 & 0 & -\mu \\ 2 & 0 & 0 & 0 \\ 0 & 2 & 0 & 0 \end{pmatrix}. \tag{4.3.6}$$

The eigenvalues of $D_w f(w_0)$ are $\pm \tfrac{1}{2} \sqrt{ai}$, both with multiplicity two. Because none of these eigenvalues equals zero, we have now shown that the strongly quenched solution exists. However, because all the eigenvalues have real parts equal to zero, we have no information on the stability of this solution. This requires higher-order averaging and taking into account the $\mathcal{O}(\varepsilon)$ terms in Eqs. (4.3.1), which we have neglected so far. In Ruijgrok (1995) these (long) calculations have been performed. They show that the stability of this solution depends in a rather complicated way on the detunings and the damping coefficients. It is also shown that, in the averaged equation, the loss of stability occurs in a Hopf bifurcation.

As is well known, a Hopf bifurcation can be described by a radial variable r and an angular variable θ. The equations for r and θ have

the form
$$r' = \lambda r - br^3 + \cdots +,$$
$$\theta' = \tfrac{1}{2}\sqrt{a} + \cdots +. \qquad (4.3.7)$$

The coefficient b determines whether the Hopf bifurcation is subcritical (when $b < 0$), supercritical ($b > 0$), or degenerate ($b = 0$). The coefficient b can, in principle, be expanded in terms of $\sqrt{\varepsilon}$: $b = b_0 + \sqrt{\varepsilon}b_1 + \cdots +$. The bifurcation parameter λ, which determines the stability of the fixed point, is the above-mentioned (complicated) function of the physical parameters such as damping and detuning. We can, for instance, fix all the parameters except σ_1. Varying σ_1, we find that there is a critical value $\sigma_1 = \sigma_c$ such that $\lambda = 0$, at which point the system undergoes the Hopf bifurcation.

In this case, we have a special situation: The coefficient b turns out to be of $\mathcal{O}(\sqrt{\varepsilon})$ (i.e., $b_0 = 0$). In other words, to first order in $\sqrt{\varepsilon}$ the Hopf bifurcation is degenerate. This special situation is called a quasi-degenerate Hopf bifurcation. Because the square of the amplitude of the periodic solution associated with a Hopf bifurcation (in the case of a supercritical bifurcation) is given by $r^2 = (\lambda/b)$, we see that, for a quasi-degenerate bifurcation, this expression is of $\mathcal{O}(\varepsilon^{-\frac{1}{2}}\lambda)$. In the case of a normal Hopf bifurcation, this would be of $\mathcal{O}(\lambda)$.

It follows from the previous considerations that the periodic solution that emerges from the Hopf bifurcation of the strongly quenched solution grows very rapidly when a bifurcation parameter, say σ_1, passes the critical point. This has also been observed in numerical experiments: When the strongly quenched solution loses stability, the x component grows very rapidly. Thus the strongly quenched solution is very sensitive to (almost) exact tuning. When detuning becomes too large, it is quickly replaced with a large-amplitude solution.

4.4 Large-Scale Motion of the Pendulum

Nearly all the literature on the mass–spring–pendulum problem has been restricted to limited vibrations of the pendulum near $\varphi = 0$. In this section we consider the upright equilibrium position and large deviations of the pendulum that result in transient processes and pendulum rotation. This is an interesting problem field with still many unsolved questions. Most of the results presented here are based on the work of Tondl et al. (1996).

4.4.1 The Semitrivial Solution $\varphi = \pi$ and Its Stability

Is it possible for the upright equilibrium position of the pendulum, corresponding with $\varphi = \pi$, to be stable? To consider this problem we replace in Eqs. (4.1.4) the angle φ with $\varphi + \pi$ to obtain the system

$$x'' + \varepsilon \kappa_1 x' + \frac{\omega_1^2}{\omega^2} x - \mu(\varphi'' \sin \varphi + \varphi'^2 \cos \varphi) = a \cos \tau,$$

$$\varphi'' + \varepsilon \kappa_2 \varphi' - \frac{\omega_0^2}{\omega^2} \sin \varphi - x'' \sin \varphi = 0. \qquad (4.4.1)$$

At this stage we have scaled only the damping coefficients, which are supposed to be small. Linearisation around the semitrivial solution, as in Subsection 3.1.3, yields the system

$$u'' + \varepsilon \kappa_1 u' + \frac{\omega_1^2}{\omega^2} u = 0,$$

$$v'' + \varepsilon \kappa_2 v' + \left[-\frac{\omega_0^2}{\omega^2} + R_0 \cos(\tau + \psi_0) \right] v = 0. \qquad (4.4.2)$$

As before we find that $u \to 0$ as $t \to \infty$, so the stability is determined by the second (Mathieu) equation, which, after a time translation, may be written as

$$v'' + \varepsilon \kappa_2 v' + (\lambda + R_0 \cos \tau) v = 0, \qquad (4.4.3)$$

where $\lambda = -(\omega_0^2/\omega^2)$. From the stability diagram of the Mathieu equation (see Figure 4.5), it follows that there indeed exist, for a given R_0, negative values of λ for which the solution $v = 0$ (corresponding to the upright position) is stable, e.g., point A. Note that for small R_0, $|\lambda|$ has to be small, which means that $\omega \gg \omega_0$: The forcing frequency of the mass–spring has to be much larger than the linear pendulum frequency.

4.4.2 Nontrivial Solutions

Quite generally the conditions for stability of the semitrivial solution with the pendulum in the upright equilibrium position will not be met. We do not have a complete picture of what happens if large-scale motion of the pendulum is possible; there are still many open problems. Here we report on a number of numerical experiments.

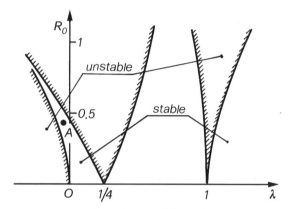

Figure 4.5: Stability diagram of the Mathieu equation.

The starting point for the experiments is again system (4.1.4), written in the following form:

$$x'' + \kappa x' + q^2 x + \mu(\varphi'' \sin \varphi + \varphi'^2 \cos \varphi) = a \cos \eta \tau,$$
$$\varphi'' + \kappa_0 \varphi' + \sin \varphi + x'' \sin \varphi = 0. \qquad (4.4.4)$$

There are four state variables, a time-periodic forcing, and various parameters, so this poses a serious problem in displaying numerical results. One possibility is to consider the time-periodic (period T of the forcing) Poincaré mapping, i.e., we plot the values of the four state variables after each interval of time T. As this picture is four dimensional it is then necessary to project this picture on a plane, which does not facilitate the interpretation.

Another possibility is to extract a special time series from the numerical data, for instance, by recording the extremal values of a certain deflection or its velocity. This is in particular interesting in quenching problems in which we are looking for a reduction of extremal values of certain variables. For instance, in our problem we are interested in the efficiency of the rotating pendulum in quenching the vibrations (the x coordinate) of the mass–spring system.

We now describe some of the numerical experiments. We present the vibration records of $x(t)$ and $\varphi(t)$; the corresponding extreme values are denoted by $[x]$, $[\varphi]$, and $[\varphi']$. The efficiency of the quenching is measured by the ratio $E_q = [\bar{x}]/R_0$, where $[\bar{x}]$ is the mean value of $[x]$ and R_0 is the vibration amplitude in the case of a nonoscillating

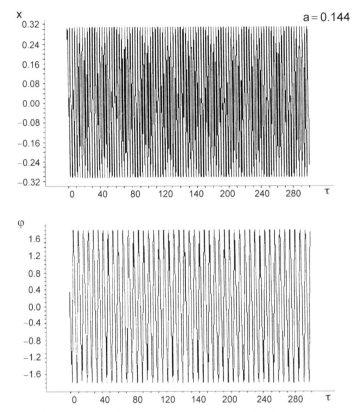

Figure 4.6: Vibration records of $x(t), \varphi(t)$ for $\kappa = 0.1, \kappa_0 = \mu = 0.05, q = 2, \eta = 2, a = 0.144$.

pendulum (the semitrivial solution). The efficiency of quenching E_q depends on not only the excitation amplitude a, but also on parameters like the damping coefficients.

We note that in Figure 4.6 the vibration is regular with frequency ratio 2:1 for $x(t)$ and $\varphi(t)$. Next we consider a case in which the motion of the pendulum is an irregular mixture of nonperiodic vibration and nonuniform rotation. Both directions of rotation are possible, and the motion appears to be chaotic. See Figure 4.7, which corresponds to the case $a = 0.2$. When a is further increased, rotation of the pendulum is typical, as illustrated in the same figure, now for the case $a = 0.325$.

Increasing the excitation amplitude a still further produces something unexpected: again nonuniform rotation may arise. We illustrate this in Figure 4.8 for $a = 0.500$.

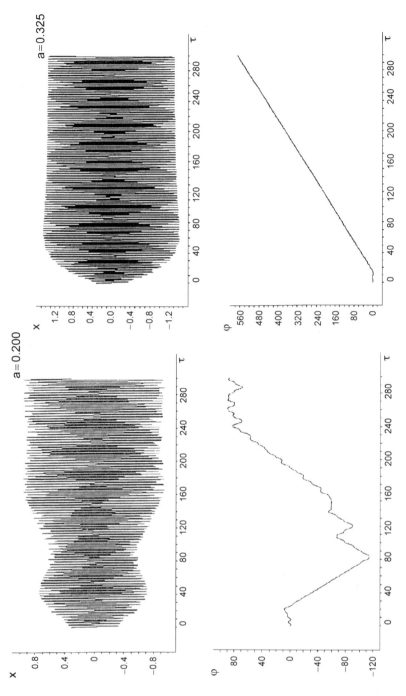

Figure 4.7: Vibration records of $x(t)$, $\varphi(t)$ corresponding to $a=0.2$ and $a=0.325$, respectively. The values of the other parameters are

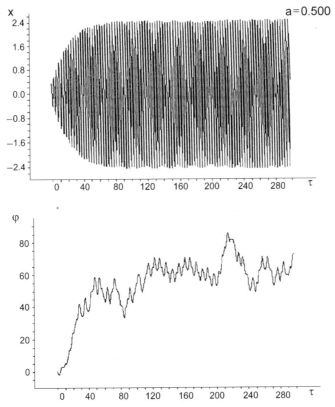

Figure 4.8: Vibration records of x(t), φ(t), for $\kappa = 0.1$, $\kappa_0 = \mu = 0.05$, $q = 2$, $\eta = 2$, $a = 0.500$.

We demonstrate the quenching efficiency E_q parameterised by the excitation amplitude a in Figure 4.9.

4.4.3 Conclusions

From these calculations we conclude that the characteristics of different motions are as follows:

1. The mass driven by the spring is vibrating with the excitation frequency while the pendulum is moving periodically or quasi-periodically with half the excitation frequency. See Figure 4.6.
2. The motion of the mass is irregular; the motion of the pendulum is an irregular mixture of nonperiodic vibration and rotation. Both

Mass–Spring–Pendulum Systems

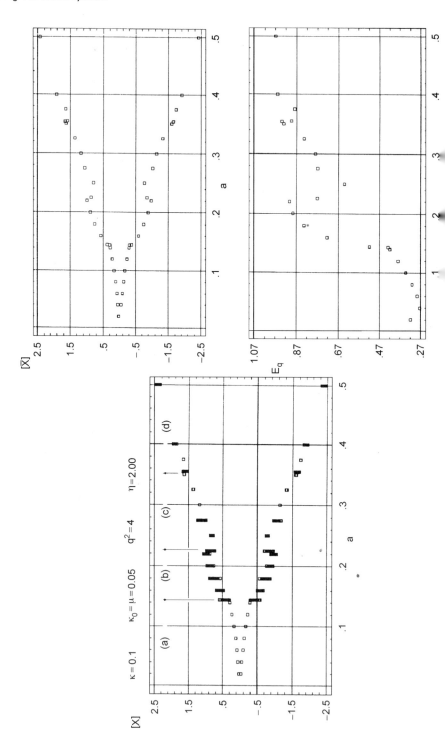

directions of rotation are possible; the motion appears to be chaotic. See the left-hand figure ($a = 0.2$) of 4.7.
3. Rotation of the pendulum is typical where the average value of the rotation frequency equals the excitation frequency of the spring. See the right-hand figure ($a = 0.325$) of 4.7.
4. Higher values of a produce again the irregular type of motion 2. See Figure 4.8.
5. As in the case of small vibrations of the pendulum, decreasing the damping of the pendulum increases the quenching efficiency. In this respect the results are similar to those for tuned absorbers.

Chapter 5

Models with More Pendulums

5.1 Introduction

In this chapter we consider systems of masses connected with springs in a chain, in which the masses can move in a vertical direction only and are periodically excited. A pendulum is attached to one or several of the masses. These systems are also examples of autoparametric systems, and resonance can be initiated if certain conditions on the natural frequencies and the driving frequency are met.

Although such systems are of considerable practical interest, the investigation of these systems has not received much attention in the literature. In fact, only the system with one-mass–spring subsystem and one pendulum has been analysed; see, e.g., Nayfeh and Mook (1979), Schmidt and Tondl (1986), and Cartmell (1990). Most authors have considered only a special tuning, namely, when the natural frequency of the pendulum is one half of the natural frequency of the secondary system (represented by the total mass mounted on the spring) and the excitation frequency is close to the latter natural frequency.

In Tondl and Nabergoj (1990) and Tondl (1992) it was shown that autoparametric resonance can occur for other values of the frequencies. This was shown by a study of the stability of the semitrivial solution, representing the vibration of the system when the pendulum does not oscillate. In Svoboda et al. (1994) this stability investigation of the semitrivial

Formulation of the Problem

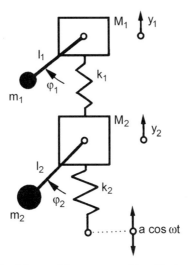

Figure 5.1: Scheme of the system considered in the example.

solution is carried out in the case in which there is just one pendulum attached to the system of masses. As an example, a two-mass system (in which the pendulum is attached to the upper mass) with a harmonic excitation acting on the upper or the lower mass was considered. The case with several masses and springs was also considered by Banerjee et al. (1996), who studied these problems by averaging. In their paper the nontrivial solutions were also studied numerically. It was found that the nontrivial solutions may lose stability by Hopf bifurcation with indications of the presence of chaotic motion. Note that this fits in nicely with the behaviour of the system considered in Chapter 4.

In this chapter, the case of several pendulum subsystems is considered; see Tondl and Nabergoj (1995) for more details. The investigation is limited to the first step only, i.e., to the analysis of the semitrivial solution. A different approach from the one in Svoboda et al. (1994) is used.

5.2 Formulation of the Problem

Let us consider a chain of N masses $M_1, M_2, \ldots, M_N$ connected with linear springs having stiffnesses $k_1, k_2, \ldots, k_N$ (see Figure 5.1). These masses can move in only the vertical direction, and their motion is damped by linear viscous damping. As in the preceding chapters, we assume that the damping is small and scale it through the small parameter ε. Some of the masses M_i ($i = 1, 2, \ldots, N$) are supporting simple pendulums, characterised by masses m_j and lengths l_j ($j = 1, 2, \ldots, n$), where $1 \leq n \leq N$. Let us denote the coordinates of the masses M_i as y_i and the angular deflections of the pendulums as φ_j. A harmonic or periodic excitation acts either directly on some of the masses $M_1, M_2, \ldots, M_N$ or by the prescribed motion of the support to which the last mass M_N is connected by a spring that has stiffness k_N.

The whole system is governed by a set of differential equations that can be divided in the following way:

- $(N - n)$ nonhomogeneous linear equations containing only the coordinates y_i, corresponding to the motion of the masses that do not have a pendulum attached
- n nonlinear equations, containing the coordinates y_i, y_j, and φ_j, corresponding to the motion of the masses that have a pendulum attached
- n nonlinear equations, containing the coordinates y_i, y_j, and φ_j, corresponding to the motion of the pendulums

These last two groups of equations have the following form:

$$(M_j + m_j)\ddot{y}_j + \varepsilon[b_j \dot{y}_j - b_{j-1,j}(\dot{y}_{j-1} - \dot{y}_j) + b_{j,j+1}(\dot{y}_j - \dot{y}_{j+1})]$$
$$- k_{j-1}(y_{j-1} - y_j) + k_j(y_j - y_{j+1})$$
$$+ m_j l_j (\ddot{\varphi}_j \sin \varphi_j + \dot{\varphi}_j^2 \cos \varphi_j) = P_j(\omega t),$$
$$m_j l_j^2 \ddot{\varphi}_j + \varepsilon c_j \dot{\varphi}_j + m_j l_j (g + \ddot{y}_j) \sin \varphi_j = 0, \quad (j = 1, 2, \ldots, n). \tag{5.2.1}$$

We note that, in the equation for y_j, the nonlinearity arises from the coupling with the pendulum φ_j. The equation for φ_j is coupled with the motion of only the jth oscillator y_j.

The semitrivial solution is defined through the excited motion of the masses $M_1, M_2, \ldots, M_N$ when the pendulums are not oscillating, i.e., $\varphi_j = 0$ ($j = 1, 2, \ldots, n$). For $\varphi_j = 0$ the nonlinear coupling terms in Eq. (5.2.1) equal zero and the semitrivial motion is therefore the steady-state solution of N linear differential equations:

$$\mathbf{M\ddot{Y}} + \varepsilon \mathbf{D\dot{Y}} + \mathbf{KY} = \mathbf{P}(\omega t). \tag{5.2.2}$$

Here **M** is the mass matrix (the mass m_j has been added to M_j), **D** is the damping matrix, **K** is the stiffness matrix, **Y** is the column vector of deflections, and $\mathbf{P}(\omega t)$ is the excitation vector.

The solution of Eq. (5.2.2), i.e., $\mathbf{Y}(\omega t)$, has the same period as $\mathbf{P}(\omega t)$. For each harmonic component of $\mathbf{P}(\omega t)$ there exists a corresponding response component. The amplitude of this component is a function of the frequency ω and has a maximum when the value of ω equals one of the eigenvalues of system (5.2.2), i.e., in the case of external resonance.

5.3 Stability of the Semitrivial Solution

We now investigate the stability of the semitrivial solution. Inserting into system (5.2.1) the perturbative solutions $y_i = y_{0i} + v_i$ and $\varphi_j = \varphi_{0j} + \psi_j$, where v_i and ψ_j are perturbations and $y_{0i} \neq 0$, $\varphi_{0j} = 0$ represents the semitrivial solution, we obtain the following differential equations in the variations:

$$\mathbf{M}\ddot{\mathbf{V}} + \varepsilon \mathbf{D}\dot{\mathbf{V}} + \mathbf{K}\mathbf{V} = 0,$$

$$m_j l_j^2 \ddot{\psi}_j + \varepsilon c_j \dot{\psi}_j + m_j l_j [g + \ddot{y}_{0j}(\omega t)]\psi_j = 0, \quad (j = 1, 2, \ldots, n),$$
(5.3.1)

where **V** is the column vector of the perturbations v_i. Assuming positive damping in the whole system, it follows that $\lim_{t \to \infty} v_i = 0$, i.e., the stability of the semitrivial solution is determined by only the second set of equations (5.3.1). Their number equals the number of pendulums, they are not mutually coupled, and they can be analysed separately. These equations are, in general, Hill equations. In the case of harmonic excitation, the response $y_{0j}(\omega t)$ is also harmonic, and they reduce to Mathieu equations, which were studied in Chapter 2.

The eigenfrequency of the jth pendulum is given by $Q_j = (g/l_j)^{1/2}$. From the theory of Mathieu equations, it follows that parametric resonance can be initiated when $\omega \approx 2Q_j$ for some j. Therefore, if the lengths of the pendulums are all different, at most one pendulum will become unstable for a particular value of ω.

If the connecting springs of masses M_j are linear, the above-mentioned approach gives the exact semitrivial solution. When some springs are nonlinear the same approach can be used, but the semitrivial solution must be determined by an approximation method.

5.4 An Illustrative Example

We consider as an example a system consisting of two masses that are connected by linear springs. Pendulum subsystems are attached to these masses (see Figure 5.1). For reasons of simplicity, it is assumed that a harmonic excitation is applied to the support (kinematic excitation). This system is governed by the following equations:

$$(M_1 + m_1)\ddot{z}_1 + \varepsilon b_1(\dot{z}_1 - \dot{z}_2) + k_1(z_1 - z_2)$$
$$+ m_1 l_1(\ddot{\varphi}_1 \sin\varphi_1 + \dot{\varphi}_1^2 \cos\varphi_1) = (M_1 + m_1)a\omega^2 \cos\omega t,$$

$$(M_2 + m_2)\ddot{z}_2 - \varepsilon b_1(\dot{z}_1 - \dot{z}_2) - k_1(z_1 - z_2) + \varepsilon b_2 \dot{z}_2 + k_2 z_2$$
$$+ m_2 l_2(\ddot{\varphi}_2 \sin\varphi_2 + \dot{\varphi}_2^2 \cos\varphi_2) = (M_2 + m_2)a\omega^2 \cos\omega t,$$

$$m_j l_j^2 \ddot{\varphi}_j + \varepsilon c_j \dot{\varphi}_j + m_j l_j g \sin\varphi_j + m_j l_j (\ddot{z}_j - a\omega^2 \cos\omega t)\sin\varphi_j = 0,$$

where $j = 1, 2$. Here z_1 and z_2 are the coordinates of the masses, relative to the lower spring end, which has a prescribed motion with amplitude a and frequency ω. The absolute coordinates are given by $y_j = z_j + a\cos\omega t$.

The first two equations above represent the primary system, which has 2 degrees of freedom. If the secondary system is at rest, $\varphi_1 = \dot{\varphi}_1 = \varphi_2 = \dot{\varphi}_2 = 0$, and the primary system is a linear system. Using the time transformation $\tau = \omega_1 t$, we can transform the above equations into the following dimensionless form:

$$w_1'' + \varepsilon \kappa_1 (w_1' - w_2') + w_1 - w_2 + \mu_1 (\varphi_1'' \sin\varphi_1 + \varphi_1'^2 \cos\varphi_1)$$
$$= \eta^2 \cos\eta\tau,$$

$$w_2'' - \mu[\varepsilon \kappa_1 (w_1' - w_2') + w_1 - w_2] + \mu_2 (\varphi_2'' \sin\varphi_2 + \varphi_2'^2 \cos\varphi_2)$$
$$+ \varepsilon \kappa_2 w_2' + q^2 w_2 = \eta^2 \cos\eta\tau,$$

$$\varphi_j'' + \varepsilon \delta_j \varphi_j' + \left[Q_j^2 + A_j(w_j'' - \eta^2 \cos\eta\tau)\right]\sin\varphi_j = 0, \quad (5.4.1)$$

where $j = 1, 2$ and $w_j = z_j/a$, $\omega_1^2 = k_1/(M_1 + m_1)$, $\kappa_j = b_j/\omega_1(M_j + m_j)$, $\mu_j = m_j l_j/a(M_j + m_j)$, $\mu = (m_1 + M_1)/(m_2 + M_2)$, $q^2 = k_2/(M_2 + m_2)\omega_1^2$, $\delta_j = c_j/\omega_1 m_j l_j^2$, $Q_j^2 = g/\omega_1^2 l_j$, and $A_j = a/l_j$ ($j = 1, 2$).

The semitrivial solution of system (5.4.1) is the steady-state solution of the subset of differential equations:

$$\mathbf{E}\mathbf{W}'' + \varepsilon \mathbf{D}\mathbf{W}' + \mathbf{K}\mathbf{W} = \eta^2 \mathbf{P} \cos \eta \tau, \qquad (5.4.2)$$

where

$$\mathbf{E} = \begin{pmatrix} 1 & 0 \\ 0 & 1 \end{pmatrix}, \quad \mathbf{D} = \begin{pmatrix} \kappa_1 & -\kappa_1 \\ -\mu\kappa_1 & \mu\kappa_1 + \kappa_2 \end{pmatrix},$$

$$\mathbf{K} = \begin{pmatrix} 1 & -1 \\ -\mu & \mu + q^2 \end{pmatrix}, \quad \mathbf{P} = \begin{pmatrix} 1 \\ 1 \end{pmatrix}, \quad \mathbf{W} = \begin{pmatrix} w_1 \\ w_2 \end{pmatrix}.$$

According to the general theory of linear differential equations, the steady-state solution of system (5.4.2) has the form

$$w_j(\tau) = a_j \cos \eta \tau + b_j \sin \eta \tau, \quad (j = 1, 2) \qquad (5.4.3)$$

where the coefficients a_j and b_j can be determined by solving the algebraic equation

$$\begin{pmatrix} 1 - \eta^2 & \varepsilon\kappa_1\eta & -1 & -\varepsilon\kappa_1\eta \\ -\varepsilon\kappa_1\eta & 1 - \eta^2 & \varepsilon\kappa_1\eta & -1 \\ -\mu & -\mu\varepsilon\kappa_1\eta & \mu + q^2 - \eta^2 & \mu\varepsilon\kappa_1 + \varepsilon\kappa_2 \\ \mu\varepsilon\kappa_1\eta & -\mu & -(\mu\varepsilon\kappa_1 + \varepsilon\kappa_2) & \mu + q^2 - \eta^2 \end{pmatrix} \begin{pmatrix} a_1 \\ b_1 \\ a_2 \\ b_2 \end{pmatrix} = \eta^2 \begin{pmatrix} 1 \\ 0 \\ 1 \\ 0 \end{pmatrix}.$$

The question of the stability of the semitrivial solution leads, with the approach of Section 5.3, to the following perturbation equation:

$$\psi_j'' + \varepsilon \delta_j \psi_j' + \{Q_j^2 - A_j \eta^2 [(1 + a_j) \cos \eta \tau + b_j \sin \eta \tau]\} \psi_j = 0,$$
$$(j = 1, 2). \qquad (5.4.4)$$

So, in the first instability region, the stability boundary curve is determined by

$$R_j = \frac{2}{\eta^2 A_j} [(Q_j^2 - \tfrac{1}{4}\eta^2)^2 + \tfrac{1}{4}\eta^2 \delta_j^2]^{1/2}.$$

It is necessary to take into consideration that in this example the solutions w_j represent the nondimensional relative coordinates with respect to the end of the lower spring. The absolute coordinates of the masses are

$$y_j(\tau) = a[(1 + a_j) \cos \eta \tau + b_j \sin \eta \tau], \quad (j = 1, 2).$$

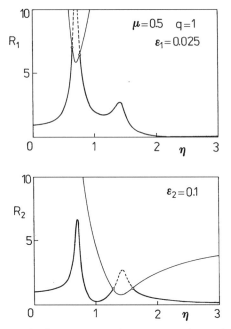

Figure 5.2: Vibration amplitude curves R_1 and R_2 corresponding to the semitrivial solution (stable solution, heavy solid curves; unstable solution, dashed curves) and the stability boundary curves R (light solid curves) as functions of the excitation frequency η. The following values have been used: $\kappa_1 = \kappa_2 = 0.10$, $\delta_1 = \delta_2 = 0.05$, $\mu = 0.5$, $q = 1.0$, $\varepsilon_1 = 0.025$, $\varepsilon_2 = 0.100$.

In all subsequent diagrams the following parameter values are used: $\kappa_1 = \kappa_2 = 0.10$, and $\delta_1 = \delta_2 = 0.05$. Different alternatives are taken for parameters μ, q, A_j, and Q_j ($j = 1, 2$). However, it is necessary to keep in mind that A_j and Q_j depend on $1/l_j$, so when a certain value of Q_j is chosen, then A_j is not arbitrary and vice versa. In all the diagrams the values of μ, q, and A_j, are marked. Because there are two pendulums attached to the masses, two diagrams are always necessary for each alternative.

Three alternatives are presented in Figures 5.2–5.4. The amplitude curves $R_1(\eta)$ and $R_2(\eta)$ are marked with heavy solid curves for stable solutions and with dashed curves for unstable solutions. Light solid curves denote again the stability boundary curves. In all the examples considered, an appropriate tuning of the pendulums has been chosen: Q_1 is close to one half of η at lower resonance, Q_2 is close to one half of η at higher resonance. We can see that for such a tuning the autoparametric

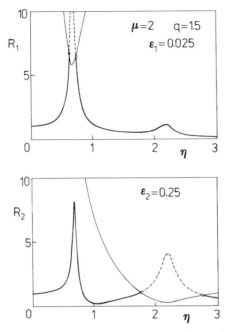

Figure 5.3: Vibration amplitude curves R_1 and R_2 corresponding to the semitrivial solution (stable solution, heavy solid curves; unstable solution, dashed curves) and the stability boundary curves R (light solid curves) as functions of the excitation frequency η. The following values have been used: $\kappa_1 = \kappa_2 = 0.10$, $\delta_1 = \delta_2 = 0.05$, $\mu = 2.0$, $q = 1.5$, $\varepsilon_1 = 0.025$, $\varepsilon_2 = 0.250$.

resonance can be initiated at a certain resonance of the primary system because of the action of only one single pendulum. This means that, at a certain resonance of the primary system, part of the excitation energy will flow into the pendulum subsystem, which results in diminishing the vibration amplitudes of the primary system.

5.5 Conclusions

The approach presented here, which uses the stability investigation of the semitrivial solution, enables us to find the optimal tuning of pendulum subsystems attached to the masses of the oscillator, consisting of several masses and springs arranged in a chain. At a convenient tuning of a pendulum (the natural frequency of the pendulum subsystem is close to one half of the excitation frequency at a certain resonance) the mass

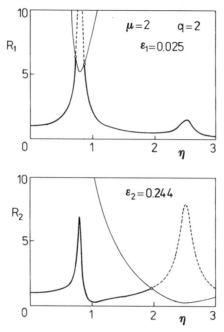

Figure 5.4: Vibration amplitude curves R_1 and R_2 corresponding to the semitrivial solution (stable solution, heavy solid curves; unstable solution, dashed curves) and the stability boundary curves R (light solid curves) as functions of the excitation frequency η. The following values have been used: $\kappa_1 = \kappa_2 = 0.10$, $\delta_1 = \delta_2 = 0.05$, $\mu = 2.0$, $q = 2.0$, $\varepsilon_1 = 0.025$, $\varepsilon_2 = 0.244$.

subsystem with pendulums can diminish the vibration amplitudes because of the flow of a part of the primary system energy into the pendulum subsystem. In most cases only one resonance of the primary system can be influenced with one pendulum.

Chapter 6

Ship Models

6.1 Introduction

Many authors have carried out mechanical or computer simulations of ship motions by using corresponding models with different levels of sophistication. The study of roll motion, i.e., the ship motion with the maximum degree of nonlinearity, is well suited to be tackled by these methods, as they offer the possibility of simulating free roll oscillations, forced rolling in a beam sea, and also the parametrically excited rolling in longitudinal or oblique seas. See Figure 6.1 for nomenclature.

The principal advantage of this equivalent approach is associated with the simplicity of handling mechanical systems or computer models compared with the difficulties encountered in both full-scale and model tests, thus allowing an easier and quicker analysis of a large number of different conditions for the ship. The simulation has certainly contributed to a better understanding of many unusual phenomena connected with ship motion as well as with the formulation of improved mathematical models used for practical evaluations.

Among the problems partially unsolved in ship dynamics, even for the case of regular waves, we would like to understand completely and exhaustively the phenomenon of parametrically excited rolling. Indeed, a longitudinal or oblique sea can yield parametrically excited transversal oscillations, which may rapidly increase their amplitude and, in some

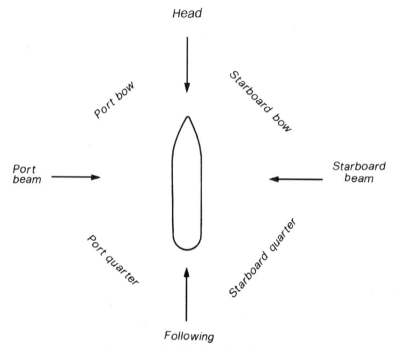

Figure 6.1: Classification of ship headings.

cases, lead to the capsizing of the vessel. The phenomenon may be established with different mechanisms that, at least in a first approximation, can be summarised as follows. The first one [Grim (1952), Kerwin (1955)] is due to a periodic change in the metacentric height of the vessel as a consequence of the travelling of the wave along the ship. In principle, it is equivalent to the parametrically induced swings of a pendulum when the pivoting point is oscillating up and down. This aspect is well known and therefore is not considered here. The second and the third mechanisms are less known [Paulling and Rosenberg (1959), Paulling (1961)] and may result from a nonlinear coupling between heave–roll or pitch–roll motions, respectively. The coupling becomes more and more effective close to the resonant conditions for the vertical motions, thus allowing an energy transfer to the transversal motion [Nayfeh et al. (1973), Mook et al. (1974)]. Its characteristic features are rather complex and not well established [Tondl and Nabergoj (1990, 1992)].

In this chapter, two different models are presented and analysed in detail, and it is shown that they are suitable for simulating a parametrically excited ship rolling in a regular, longitudinal, or oblique sea. The coupled heave–roll motion is described by means of the first model, whereas the second one is suitable for modelling the more complex case of heave–pitch–roll motion. Such simple models cannot reproduce in great detail the very complex phenomenon of ship–fluid interaction, but they are physically correct insofar as both the restoring and the coupling effects, which constitute the governing terms in the dynamics of the vessel, are considered. The identification of the ship–oscillator correspondences in the previous models is straightforward and therefore is not considered in detail.

6.2 Simplified Model for Heave–Roll Motion

The mechanical system considered is schematically represented in Figure 1.6. It consists of a mass restrained by an elastic spring that, in turn, carries a simple pendulum made up of a mass attached to a hinged weightless rod. The system is forced to oscillate sinusoidally in the vertical direction by means of an external driver with constant amplitude and frequency.

This system is able to simulate the dynamic behaviour of a vessel running in a regular longitudinal or oblique sea and therefore gives the possibility of reproducing the nonlinear coupling between heaving and rolling motions. It is easy to understand that the vertical motion of the mass corresponds to heave and that the motion of the pendulum corresponds to roll. The coupling among the oscillations is accomplished by connecting the two masses, and the effect of the waves is simulated by means of external forcing. The mechanical characteristics are chosen so as to reproduce the natural frequencies of both ship motions, the damping and the hydrodynamic reactions, and the sea effect.

The motion of the system is to be considered, and its governing equations, by introducing a vertical axis pointing upwards, are

$$(M+m)(\ddot{z} - \alpha\omega^2 \cos \omega t) + b\dot{z} + kz + ml(\ddot{\varphi} \sin \varphi + \dot{\varphi}^2 \cos \varphi) = 0,$$

$$ml^2\ddot{\varphi} + c\dot{\varphi} + mgl \sin \varphi + ml(\ddot{z} - \alpha\omega^2 \cos \omega t) \sin \varphi = 0,$$

$$(6.2.1)$$

where z is the relative vertical displacement of the mass M, φ is the angular displacement of the mass m, k is the elastic constant of the spring, l is the length of the rod, b and c are the damping coefficients of the linear and the angular motions, respectively, and α and ω are the amplitude and the frequency of the excitation, respectively. The overdot indicates the derivative with respect to time t. Equations (6.2.1) represent a system of two coupled nonlinear differential equations of second order, and by means of the time variable $\tau = (g/l)^{1/2} t$, they can be written in the following dimensionless form:

$$w'' + \kappa w' + q^2 w + \mu(\varphi'' \sin\varphi + \varphi'^2 \cos\varphi) = a\eta^2 \cos\eta\tau,$$
$$\varphi'' + \kappa_0 \varphi' + \sin\varphi + (w'' - \varepsilon\eta^2 \cos\eta\tau)\sin\varphi = 0, \qquad (6.2.2)$$

where $w = z/l$, $\omega_0 = (g/l)^{1/2}$, $\kappa = b/\omega_0(M+m)$, $q^2 = k/\omega_0^2(M+m)$, $\mu = m/(M+m)$, $\kappa_0 = c/\omega_0 m l^2$, $\eta = \omega/\omega_0$, $a = \alpha/l$, and the prime indicates the derivative with respect to the new time variable. The steady-state solution of system (6.2.2) is given by

$$w_0(\tau) = a(A\cos\eta\tau + B\sin\eta\tau),$$
$$\varphi_0(\tau) = 0, \qquad (6.2.3)$$

where $A = \eta^2(q^2 - \eta^2)/\Delta$, $B = \kappa\eta^3/\Delta$, $\Delta = (q^2 - \eta^2)^2 + (\kappa\eta)^2$.

We can determine the stability of the steady-state solutions by superposing small perturbations on system (6.2.3), that is, by letting

$$w = w_0 + u,$$
$$\varphi = \varphi_0 + \psi, \qquad (6.2.4)$$

and by substituting Eqs. (6.2.4) into Eqs. (6.2.2). Consequently, at a first order of approximation, the resulting perturbation equations are

$$u'' + \kappa u' + q^2 u = 0,$$
$$\psi'' + \kappa_0 \psi' + \psi - a\eta^2[(1+A)\cos\eta\tau + B\sin\eta\tau]\psi = 0. \qquad (6.2.5)$$

The first equation of system (6.2.5) is a homogeneous linear differential equation with constant coefficients and its solution is asymptotically stable. The second equation is a Mathieu equation that may have stable or unstable solutions, depending on the values taken for the different parameters.

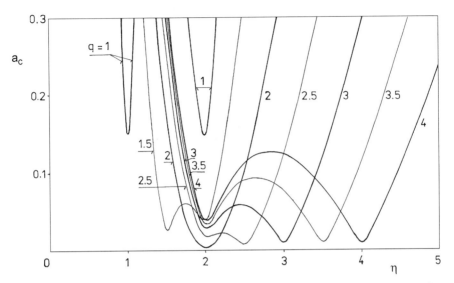

Figure 6.2: Instability threshold a_c corresponding to the semitrivial solution as a function of the excitation frequency η for different values of tuning ratio q. The instability region is under the curves. The following values have been used: $\kappa_0 = \kappa = 0.10$.

In the first instability region, where the solution of the Mathieu equation may be approximated in the form

$$\psi(\tau) = C \cos \tfrac{1}{2}\eta\tau + D \sin \tfrac{1}{2}\eta\tau, \qquad (6.2.6)$$

we can obtain the limiting curve, which separates the stability region from the instability region, by inserting Eq. (6.2.6) into the second equation of system (6.2.5). The nontrivial solution for C and D results in

$$\left(1 - \tfrac{1}{4}\eta^2\right)^2 + \tfrac{1}{4}\kappa_0^2\eta^2 - \tfrac{1}{4}a^2\eta^4[(1+A)^2 + B^2] = 0. \qquad (6.2.7)$$

After simple calculations, the instability threshold for the appearance of a parametric resonance may be expressed in the following explicit form:

$$a = \frac{2}{\eta^2}\left[\frac{\left(1 - \tfrac{1}{4}\eta^2\right)^2 + \tfrac{1}{4}\kappa_0^2\eta^2}{(1+A)^2 + B^2}\right]^{1/2}. \qquad (6.2.8)$$

As an example, in Figure 6.2 the instability threshold $a_c = a_c(\eta)$ is depicted for $\kappa_0 = \kappa = 0.10$ and for different values of q, the synchronisation ratio. The instability region is present within the upper part of the curves for each value of q. It may be observed that, unless q is not

Ship Models

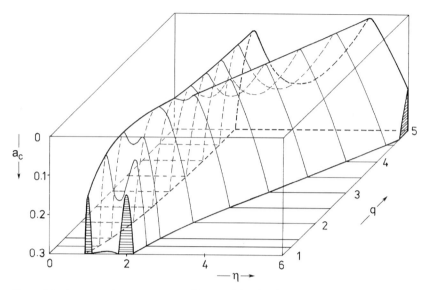

Figure 6.3: Axonometric representation of the instability threshold $a_c = a_c(\eta, q)$ of the semitrivial solution. The instability region is under the surface. The following values have been used: $\kappa_0 = \kappa = 0.10$.

close to 2, the threshold exhibits two minima, one for η close to the value 2 and the other for η close to the value q. For $q = 2$, the minima will merge and a single minimum will appear for $\eta = 2$. The geometric characteristics of the limiting curves can be better visualised if the instability threshold is represented by means of an axonometric system of the form $a_c = a_c(\eta, q)$, i.e., by a three-dimensional surface. In Figures 6.3 and 6.4 the threshold surface $a_c = a_c(\eta, q)$ is depicted for $\kappa_0 = \kappa = 0.10$, and $\kappa_0 = 0.10$ and $\kappa = 0.20$, respectively. For convenience of the representation, we have changed the positive direction of the a axis by orienting it downwards. In this way the instability region is located under the surface and the minima appear as maxima. An increase of κ, the damping coefficient of the vertical motion determines a growth of the minima for $\eta \approx q$, which corresponds to a decrease of the maxima in Figure 6.4.

The possibility of the appearance of a parametric resonance in the rolling motion of a vessel is greatly increased if the instability threshold presents a minimum. Thus, in a regular longitudinal or oblique sea, this will occur when the frequency ratio between the excitation and the

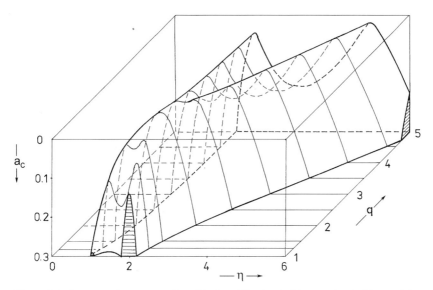

Figure 6.4: Axonometric representation of the instability threshold $a_c = a_c(\eta, q)$ of the semitrivial solution. The instability region is under the surface. The following values have been used: $\kappa_0 = 0.10$, $\kappa = 0.20$.

rolling equals two and, also, when the heave motion becomes resonant, i.e., for $\eta \approx q$. This is certainly an important aspect of the nature of the heave–roll coupling phenomenon, which is controlled by both roll and heave damping. An increase of the damping may produce a drastic decrease of the possibility of parametric resonance and thus may reduce the risk of dangerous transversal oscillations for the ship.

6.3 Effect of the Spring Nonlinearity

In Section 6.2, a linear characteristic for the buoyancy force acting on the vessel has been considered. For real ships the characteristic is nonlinear and nonsymmetric. Thus, if the spring nonlinearity is expressed by a second-order term in the displacement, the resulting system is governed by the following differential equations:

$$w'' + \kappa w' + q^2(1 - \gamma w)w + \mu(\varphi'' \sin \varphi + \varphi'^2 \cos \varphi) = a\eta^2 \cos \eta\tau,$$
$$\varphi'' + \kappa_0 \varphi' + \sin \varphi + (w'' - a\eta^2 \cos \eta\tau) \sin \varphi = 0. \qquad (6.3.1)$$

Equations (6.3.1) admit the semitrivial solution, which can be approximated in the form

$$w_0(\tau) = a(X + A\cos\eta\tau + B\sin\eta\tau).$$
$$\varphi_0(\tau) = 0. \tag{6.3.2}$$

After substituting Eqs. (6.3.2) into Eqs. (6.3.1) and by using the method of harmonic balance, we obtain the following equations:

$$X - \gamma\left(X + \tfrac{1}{2}R_0^2\right) = 0,$$
$$[q^2(1 - 2\gamma X) - \eta^2]A + \kappa\eta B = \eta^2,$$
$$-\kappa\eta A + [q^2(1 - 2\gamma X) - \eta^2]B = 0, \tag{6.3.3}$$

where $R_0 = (A^2 + B^2)^{1/2}$ is the amplitude of the oscillation. From the first equation of Eqs. (6.3.3) it follows that

$$R_0 = 2X\left(\frac{1}{\gamma} - X\right). \tag{6.3.4}$$

The second and the third equations of Eqs. (6.3.3) yield the relations

$$A = \eta^2(g^2 - \eta^2)/\Delta, \quad B = \kappa\eta^3/\Delta, \quad \text{where } g^2 = q^2(1 - 2\gamma X),$$
$$\Delta = (g^2 - \eta^2)^2 + (\kappa\eta)^2.$$

The frequency response curve $R_0 = R_0(\eta)$ can be obtained from the relation

$$R_0^2 = A^2 + B^2 = \frac{\eta^4}{(g^2 - \eta^2)^2 + (\kappa\eta)^2}, \tag{6.3.5}$$

which can be easily rearranged into the form

$$\left(1 - \frac{1}{R_0^2}\right)\eta_{1,2}^2 = g^2 - \tfrac{1}{2}\kappa^2 \pm \left(\frac{g^4}{R_0^2} - \kappa^2 g^2 + \tfrac{1}{4}\kappa^4\right)^{1/2}. \tag{6.3.6}$$

The following strategy can be used for calculating the dependences $R_0(\eta)$ and $X(\eta)$. The value of X is changed step by step, the value of R_0 is calculated from Eq. (6.3.4) and the corresponding value of η from Eq. (6.3.6).

The stability of the semitrivial solution can be determined by the analysis of the differential equations of the disturbed motion, which we obtain by inserting

$$w = w_0 + u,$$
$$\varphi = \varphi_0 + \psi, \qquad (6.3.7)$$

into Eqs. (6.3.1):

$$u'' + \kappa u' + q^2(1 - 2\gamma w_0)u = 0,$$
$$\psi'' + \kappa_0 \psi' + \psi - a\eta^2 [(1 + A)\cos \eta\tau + B \sin \eta\tau]\psi = 0. \qquad (6.3.8)$$

The first perturbative equation is identical with that obtained when the stability of the system with a nonoscillating pendulum was investigated. It was proved by Tondl et al. (1970) that the rule of vertical tangents is valid for this particular system with 1 degree of freedom: In the interval of the excitation frequency where three steady-state solutions exist, the solutions with the highest and the lowest amplitudes are stable, the middle one is unstable. Thus, because Eqs. (6.3.8) are not mutually coupled, for the stability of the semitrivial solution the second equation is decisive.

Approximating the solution on the boundary of the main instability domain in the form

$$\psi(\tau) = C \cos \tfrac{1}{2}\eta\tau + D \sin \tfrac{1}{2}\eta\tau, \qquad (6.3.9)$$

and, after substituting Eq. (6.3.9) into Eqs. (6.3.8), we obtain the following equations:

$$\left[1 - \tfrac{1}{4}\eta^2 - \tfrac{1}{2}a\eta^2(1+A)\right]C + \left(\tfrac{1}{2}\kappa_0\eta - \tfrac{1}{2}a\eta^2 B\right)D = 0,$$
$$-\left(\tfrac{1}{2}\kappa_0\eta + \tfrac{1}{2}a\eta^2 B\right)C + \left[1 - \tfrac{1}{4}\eta^2 + \tfrac{1}{2}a\eta^2(1+A)\right]D = 0. \qquad (6.3.10)$$

The condition of the nontrivial solution of C and D leads to the relation

$$\left(1 - \tfrac{1}{4}\eta^2\right)^2 + \tfrac{1}{4}\kappa_0^2\eta^2 - \tfrac{1}{4}a^2\eta^4[(1+A)^2 + B^2] = 0, \qquad (6.3.11)$$

which is identical to Eq. (6.2.7).

Here, a different approach from that shown in Section 6.2 is presented in which we consider the dimensionless amplitude of the absolute motion:

$$(R_a)_0^2 = (1+A)^2 + B^2 = \frac{g^4 + \kappa^2\eta^2}{(g^2 - \eta^2)^2 + (\kappa\eta)^2}. \qquad (6.3.12)$$

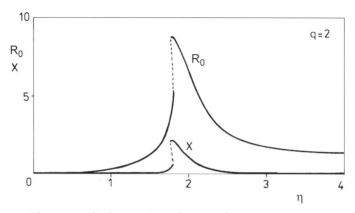

Figure 6.5: Vibration amplitude curve R_0 and constant bias curve X corresponding to the semitrivial solution (stable solution, heavy solid curves; unstable solution, dashed curves) as a function of the excitation frequency η. The following values have been used: $\kappa = 0.10$, $\gamma = 0.05$, $q = 2$.

In this way, relation (6.3.11) can be used to express the stability boundary curve. After simple rearrangement, it becomes

$$R_a = \frac{2}{a\eta^2}\left[\left(1 - \tfrac{1}{4}\eta^2\right)^2 + \tfrac{1}{4}\kappa_0^2\eta^2\right]^{1/2}. \qquad (6.3.13)$$

Thus the instability intervals are obtained by the section points of the absolute motion amplitude curve of Eq. (6.3.12) and the stability boundary curve of Eq. (6.3.13).

The dependences $R_0(\eta)$ and $X(\eta)$ for $\kappa = 0.10$, $\gamma = 0.05$, and $q = 2$ are shown in Figure 6.5. Unstable solutions according to the rule of vertical tangents are marked by dashed curves. Diagrams of $(R_a)_0(\eta)$ for different values of q and the same values of κ and γ as in the previous case, together with the stability boundary curves $R_a(\eta)$ for $\kappa_0 = 0.05$ and different values of q and a, are presented in Figure 6.6. For further illustration, the same curves are shown in Figure 6.7 for $q = 3$ and $a = 0.05$. Stable solutions are marked by solid heavy curves. We can see that for this particular case even two instability intervals of the semitrivial solution exist where the autoparametric resonance vibration is initiated.

From the preceding results, it follows that the basic properties of the two systems with linear and nonlinear springs that have nonsymmetric characteristics are very similar. As far as the stability of the semitrivial

Effect of the Spring Nonlinearity

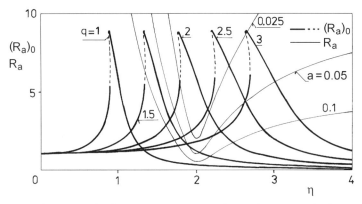

Figure 6.6: Vibration amplitude curve $(R_a)_0$ corresponding to the semitrivial solution (stable solution, heavy solid curves; unstable solution, dashed curves) and the stability boundary curve R_a (light solid curves) as a function of the excitation frequency η for different values of tuning ratio q and excitation intensity a. The following values have been used: $\kappa_0 = 0.05$, $\kappa = 0.10$, $\gamma = 0.05$.

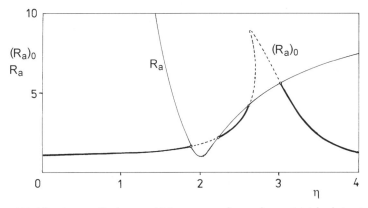

Figure 6.7: Vibration amplitude curve $(R_a)_0$ corresponding to the semitrivial solution (stable solution, heavy solid curves; unstable solution, dashed curves) and the stability boundary curve R_a (light solid curves) as a function of the excitation frequency η. The following values have been used: $\kappa_0 = 0.05$, $\kappa = 0.10$, $\gamma = 0.05$, $q = 3$, $a = 0.05$.

solution is concerned, even two instability intervals of the excitation frequency can exist for both systems ($\eta \approx 2$ and $\eta \approx q$), the latter being slightly shifted towards lesser values of η when the spring nonlinearity is increased.

Ship Models

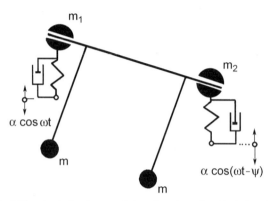

Figure 6.8: Model for simulating heave–pitch–roll motion of a ship in longitudinal waves.

6.4 Extension to Heave–Pitch–Roll Motion

The mechanical system shown in Figure 6.8 is to be considered. It consists of two masses restrained by elastic springs and supporting two equal pendulums rigidly connected by means of a weightless rod. The masses are excited at the same frequency and amplitude, but with a certain phase lag, to consider the delayed effects of the wave propagating along the ship. This system basically doubles the mechanical model considered in Section 6.2.

This mechanical system was developed to simulate the most general case of ship motion in a longitudinal or oblique sea with the possibility of a nonlinear coupling between heave–pitch–roll motions. The subdivision can be qualitatively thought of as helpful in representing the change in buoyancy along the ship that causes a pitching moment and consequently the pitching motion. Thus the motion of the two rigidly connected pendulums corresponds to roll, and the two masses in vertical motion will reproduce the simultaneous heave–pitch oscillations. When the displacements of the two masses are in phase we deal with a heaving motion, but in the case of antiphase displacements a pitching motion is obtained. In reality, none of these two motions is exactly accomplished by the system, and thus an intermediate situation of heave–pitch oscillations will usually be present.

For this particular case, the mechanical characteristics of the system must be carefully selected and their choice will be conducted in order to reproduce the natural frequencies of the ship motions of heave,

Extension to Heave–Pitch–Roll Motion

pitch, and roll, their hydrodynamic reactions and damping, and the wave excitation on the vessel as well. The relationship between the vertical displacements of the two masses and those of heave and pitch can be easily obtained together with the relationship between the corresponding natural frequencies of oscillation. This problem is not considered here; however, it is expected that if one of the masses is in resonant motion then the ship motion will also be amplified and heave or pitch will prevail. In this case, the phase lag of the vertical displacements will determine which of the motions is the prevailing one.

The equations of motion for the system shown in Figure 6.8 are

$$(M_1 + m)(\ddot{z}_1 - \alpha \omega^2 \cos \omega t) + b_1 \dot{z}_1 + k_1 z_1$$
$$+ ml(\ddot{\varphi} \sin \varphi + \dot{\varphi}^2 \cos \varphi) = 0,$$

$$(M_2 + m)[\ddot{z}_2 - \alpha \omega^2 \cos(\omega t - \psi)] + b_2 \dot{z}_2 + k_2 z_2$$
$$+ ml(\ddot{\varphi} \sin \varphi + \dot{\varphi}^2 \cos \varphi) = 0,$$

$$ml^2 \ddot{\varphi} + c\dot{\varphi} + mgl \sin \varphi + \tfrac{1}{2} ml[\ddot{z}_1 - \alpha \omega^2 \cos \omega t + \ddot{z}_2$$
$$- \alpha \omega^2 \cos(\omega t - \psi)] \sin \varphi = 0,$$

where the symbols have the same meaning as in Section 6.3. Once again this system of differential equations can be reduced to a dimensionless form by means of the new time variable $\tau = (g/l)^{1/2} t$. Thus we obtain

$$w_1'' + \kappa_1 w_1' + q_1^2 w_1 + \mu_1(\varphi'' \sin \varphi + \varphi'^2 \cos \varphi) = a\eta^2 \cos \eta \tau,$$
$$w_2'' + \kappa_2 w_2' + q_2^2 w_2 + \mu_2(\varphi'' \sin \varphi + \varphi'^2 \cos \varphi) = a\eta^2 \cos(\eta \tau - \psi),$$
$$\varphi'' + \kappa_0 \varphi' + \sin \varphi + \tfrac{1}{2}[w_1'' - a\eta^2 \cos \eta \tau + w_2''$$
$$- \varepsilon \eta^2 \cos(\eta \tau - \psi)] \sin \varphi = 0, \quad (6.4.1)$$

where $w_j = z_j/l$, $\kappa_j = b_j/\omega_0(M_j + m)$, $q_j^2 = k_j/\omega_0^2(M_j + m)$, $\mu_j = m/(M_j + m)$ for $j = 1, 2$ and $\omega_0 = (g/l)^{1/2}$, $\kappa_0 = c/\omega_0 ml^2$, $\eta = \omega/\omega_0$, and $a = \alpha/l$. The steady-state solution of system (6.4.1) is

$$w_{0j}(\tau) = a(A_j \cos \eta \tau + B_j \sin \eta \tau),$$
$$\varphi_0(\tau) = 0, \quad (6.4.2)$$

where $A_1 = \eta^2(q_1^2 - \eta^2)/\Delta_1$, $B_1 = \kappa_1 \eta^3/\Delta_1$, $\Delta_1 = (q_1^2 - \eta^2)^2 + (\kappa_1 \eta)^2$, $A_2 = \eta^2((q_2^2 - \eta^2) \cos \psi - \kappa_2 \eta \sin \psi)/\Delta_2$, $B_2 = \eta^2((q_2^2 - \eta^2) \sin \psi + \kappa_2 \eta \cos \psi)/\Delta_2$, and $\Delta_2 = (q_2^2 - \eta^2)^2 + (\kappa_2 \eta)^2$.

We can obtain the stability of solution (6.4.2) by substituting into system (6.4.1) the perturbation solutions of the form

$$w_j = w_{0j} + u_j,$$
$$\varphi = \varphi_0 + \psi. \tag{6.4.3}$$

By performing the required mathematical manipulations, we obtain the following first-order approximation equations:

$$u_j'' + \kappa_j u_j' + q_j^2 u_j = 0,$$
$$\psi'' + \kappa_0 \psi' + \psi - \tfrac{1}{2} a \eta^2 (E \cos \eta \tau + F \sin \eta \tau) \psi = 0, \tag{6.4.4}$$

where $j = 1, 2$ and $E = 1 + A_1 + A_2 + \cos \psi$, and $F = B_1 + B_2 + \sin \psi$. The first two equations of system (6.4.4) are linear homogeneous differential equations with constant coefficients and their solutions are asymptotically stable. The third equation is again a Mathieu equation for which, in the first instability region, the limiting curve is given by

$$a_c = \frac{4}{\eta^2} \left[\frac{\left(1 - \tfrac{1}{4}\eta^2\right)^2 + \tfrac{1}{4}\kappa_0^2 \eta^2}{E^2 + F^2} \right]^{1/2}. \tag{6.4.5}$$

In the cases reported, the instability threshold for the appearance of a parametric resonance will be shown only by means of the axonometric representation in space. To carry out the computations we may consider the pair of independent variables η and ψ or the pair η and q_2 with assigned values for other parameters, i.e., $\kappa_0 = \kappa_1 = \kappa_2 = 0.10$. In Figure 6.9 the threshold $a_c = a_c(\eta, \psi)$ is shown for $q_1 = 4$ and $q_2 = 3$. In general, there are three minima, and these occur for η close to the values 2, 3, and 4. The influence of the phase ψ on the threshold level is limited to regions of the minima around q_1 and q_2 but becomes substantial in the neighbourhood of $\eta = 2$, where an increase of a_c occurs for $\psi \approx \pi$, i.e., when the two masses are excited in antiphase. The same kind of instability threshold $a_c = a_c(\eta, \psi)$ is shown in Figure 6.10, but for $q_1 = 3$ and $q_2 = 2$. Because $q_2 = 2$, there are only two minima and the influence of ψ appears substantially reduced. Finally, the graph in Figure 6.11 shows a different threshold representation, i.e., the threshold $a_c = a_c(\eta, q_2)$ for $q_1 = 3$ and $\psi = 3\pi/2$. We may note three minima, excluding the case in which q_2 is close to the value 2, where two of the minima merge.

Extension to Heave–Pitch–Roll Motion

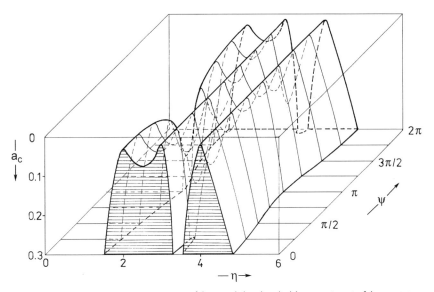

Figure 6.9: Axonometric representation of the instability threshold $a_c = a_c(\eta, \psi)$ of the semitrivial solution. The instability region is under the surface. The following values have been used: $q_1 = 4$, $q_2 = 3$, $\kappa_0 = \kappa_1 = \kappa_2 = 0.10$.

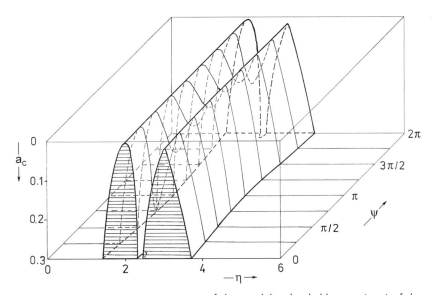

Figure 6.10: Axonometric representation of the instability threshold $a_c = a_c(\eta, \psi)$ of the semitrivial solution. The instability region is under the surface. The following values have been used: $q_1 = 3$, $q_2 = 2$, $\kappa_0 = \kappa_1 = \kappa_2 = 0.10$.

Ship Models

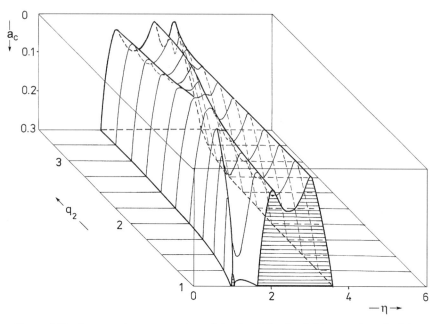

Figure 6.11: Axonometric representation of the instability threshold $a_c = a_c(\eta, q_2)$ of the semitrivial solution. The instability region is under the surface. The following values have been used: $q_1 = 3$, $\psi = 3\pi/2$, $\kappa_0 = \kappa_1 = \kappa_2 = 0.10$.

In addition to the usual resonance obtained for a frequency ratio between the excitation and the rolling equal to 2, we may find a parametric resonance each time the motion in the vertical direction of one of the two masses becomes resonant, i.e., for $\eta \approx q_1$ or $\eta \approx q_2$. As a consequence, it is expected that the phenomenon will appear when a resonance condition is fulfilled for the two corresponding motions of heave and pitch and the phase delay of the excitations has a particular role in governing the phenomenon. More precisely, we have an increase of the threshold when the masses are driven by antiphase excitations.

The choice of a complex system of two pairs of mass–pendulum subsystems corresponds, in principle, to a subdivision of the ship into two parts, the forebody and the aftbody of the ship. Such a procedure is certainly limited, but it is justified by means of the well-known strip theory commonly used in seakeeping calculations [Korvin-Kroukovsky (1961)]. Within the framework of this theory the ship is subdivided into various transversal strips, the number of which may be increased to infinity. Thus there are wide possibilities of improving this mechanical equivalence modelling by considering numerous subsystems of mass–pendulum type with rigid or elastic connections [Nabergoj and Tondl (1994)].

6.5 Conclusions

The mechanical models presented in this chapter are able to predict the onset of a parametric resonance in the rolling motion of a vessel running in a regular longitudinal or oblique sea and therefore are suitable for the simulation of both coupled heave–roll and coupled heave–pitch–roll motions. In particular, the analysis shows the occurrence of, at most, three critical frequencies for which the excitation may produce the onset of transversal oscillations. The results suggest that parametrically induced rolling may appear also when one of the vertical ship motions becomes resonant. However, only a more detailed analysis of practical examples should put into evidence which mechanism is the most important for the safety of a particular vessel.

The parametrically induced roll oscillations most likely will become more effective if the resonance frequency of heave and/or pitch is harmonic to the roll frequency. In such a case, some of the minima in the instability threshold will merge and the numbers of critical frequencies for the onset of the phenomenon will diminish from three to two or one. This fully confirms the results of early investigations of coupling between vertical and transversal motions. Indeed, it has been observed [Froude (1861)] that ships have undesirable roll characteristics if the natural frequency in heave/pitch and roll are in the ratio of 2:1 and if the roll axis does not lie in the plane of the water surface.

Chapter 7

Flow-Induced Vibrations

7.1 Introduction

Flow-induced vibrations are of major importance in many technological devices, for instance in high structures like towers, chimneys and bridges, tubes in heat exchangers, high-voltage lines, etc. In this chapter we discuss some autoparametric systems in which flow is the source of self-excitation.

It should be noted that an exact mathematical description of the forces acting on a vibrating body in a flow is very difficult and still under discussion in the literature [see, for instance, van der Burgh (1990) for the case of galloping in high-voltage lines]. The absence of a solid mathematical model is especially a problem if a nonlinear model is required for a full analysis. In engineering practise, sometimes so-called bluff-body systems are used that represent mathematical models in which the steady-state solution has the properties expressing the important features of the observed phenomena of systems in cross flow. In this way, bluff-body systems may serve as useful metaphors in the absence of solid mathematical–mechanical models. In this chapter we consider two models of this type.

One of the simplest models (the critical velocity model) of an elastically mounted body in cross flow is characterised by the following properties: when the velocity of the flow is increased, the equilibrium position is stable up to a certain limit, the critical velocity U_c. Increasing the flow velocity beyond the critical velocity results in destabilisation of the equilibrium position and initiation of self-excited vibrations of finite amplitude. Such a model can be described by the following differential equation:

$$M\ddot{y} + b\dot{y} - b_0 U^2 (1 - \gamma_0 y^2)\dot{y} + ky = 0, \qquad (7.1.1)$$

where γ_0 is the coefficient of the nonlinear term that expresses the decrease of the destabilising effect with increasing deflection of the body, M is the mass of the body, y is its deflection, k is the stiffness of the elastic mounting, b is the coefficient of positive damping, and U is the flow velocity. Moreover, $U_c = (b/b_0)^{1/2}$ is the critical velocity of the flow for which the equilibrium position of the body loses its stability. Note that the system contains positive and negative damping. For $U < U_c$ the equilibrium position is stable and any disturbance leads to decaying motion. For $U > U_c$ the equilibrium position is unstable, which leads to divergent motion in the linearised system. In real systems for $U > U_c$ a limited vibration occurs because of the action of nonlinearities. The steady-state vibration with a limited amplitude can be modelled with either the progressive damping or nonlinear dependence of the destabilising force. Vibrations arising in this way may be undesirable. We shall attach a pendulum to the vibrating mass to obtain an autoparametric system. This system is analysed in Section 7.2.

A second model (the vortex shedding model) describes the behaviour of a system in cross flow in which self-excited vibrations are induced by vortex shedding. This vortex shedding is characterised by the frequency $\omega_s = (U/D)s$, where s is the Strouhal number (incorporating the Reynolds number), U is the flow velocity, and D is the diameter of the body section. With the mass of the body again denoted by M and the stiffness of the elastic mounting by k, the natural frequency is given by $\omega_0^2 = k/M$.

The main properties of the system can be characterised by the diagrams of the vibration amplitude A and the vibration frequency Ω as depending on the vortex frequency ω_s; see Figure 7.1. We see that if $\omega_s < \omega_0$, we have $\Omega < \omega_s$, but it is very close to ω_s; the amplitude A increases with ω_s. If $\omega_s > \omega_0$, then we have $\Omega < \omega_0$, but now it is very

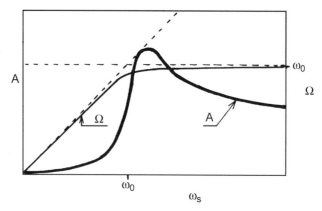

Figure 7.1: Schematic representation of the vibration amplitude A and frequency Ω of the steady-state solution as a function of the vortex shedding frequency ω_s. Here, ω_0 is the natural frequency of the system.

close to ω_0. The amplitude reaches its maximum at a value of ω_s, which is close to ω_0. However, a further increase of ω_s does not result in a rapid decrease of A, as is the case in externally excited systems, but in a slow decrease.

We now formulate a model of an elastically mounted body in cross flow that is excited because of vortex shedding, where y represents the body deflection [see Tondl (1987)]. We have the system

$$M\ddot{y} + b\dot{y} + ky = \beta_0 u,$$
$$\ddot{u} + \alpha_0 \dot{u} + \omega_s^2[u - \delta\,\mathrm{sgn}(\dot{y})] = 0, \qquad (7.1.2)$$

where the term $\beta_0 u$ is related to the lift force. The second equation of system (7.1.2) has been introduced to obtain the desired properties for the steady-state solution y. We will subsequently study the stability of the semitrivial solution and the nontrivial solutions of the critical velocity model in an autoparametric setting in Sections 7.2 and 7.3. This is followed in Section 7.4 by an extension of this model to dry friction. We also discuss the vortex shedding model in an autoparametric context in Sections 7.5 and 7.6. After this we generalise the critical velocity model and we give a detailed account of the bifurcations of the nontrivial solutions in Section 7.7.

Note that more complicated models have been proposed involving jump and hysteresis phenomena [see Tondl (1988a, 1991b)].

7.2 The Critical Velocity Model

Consider a simple one-mass model (Figure 1.9) that is excited by fluid flow according to the first model and is characterised by Eq. (7.1.1) [see Tondl and Nabergoj (1994)]. When a simple pendulum of mass m and length l is attached to the body, the two-mass system is governed by the following equations:

$$(M+m)\ddot{y} + b\dot{y} - b_0 U^2(1-\gamma_0 y^2)\dot{y} + ky$$
$$+ ml(\ddot{\varphi}\sin\varphi + \dot{\varphi}^2\cos\varphi) = 0,$$
$$ml^2\ddot{\varphi} + c\dot{\varphi} + ml(g+\ddot{y})\sin\varphi = 0, \quad (7.2.1)$$

where φ is the angular deflection, $c > 0$ is the damping coefficient, and g is the gravity acceleration. With the time transformation $\tau = \omega_1 t$, where $\omega_1 = (g/l)^{1/2}$ is the natural frequency of the pendulum, differential equations (7.2.1) can be transformed into dimensionless form:

$$x'' + \kappa x' + q^2 x - \beta V^2(1-\gamma x^2)x' + \mu(\varphi''\sin\varphi + \varphi'^2\cos\varphi) = 0,$$
$$\varphi'' + \kappa_0\varphi' + (1+x'')\sin\varphi = 0.$$
$$(7.2.2)$$

Here $x = y/l$, $\kappa = b/\omega_1(M+m)$, $q = \omega_0/\omega_1$, $\omega_0^2 = k/(M+m)$, $\beta = b_0 U_0^2/\omega_1(M+m)$, $V = U/U_0$, $\gamma = \gamma_0 l^2$, $\mu = m/(M+m)$, $\kappa_0 = c/\omega_1 ml^2$, and U_0 is a chosen value of the flow velocity.

Equations (7.2.2) admit the semitrivial solution $x_0(\tau) \neq 0$, $\varphi_0 = 0$, which satisfies

$$x'' + \kappa x' + q^2 x - \beta V^2(1-\gamma x^2)x' = 0. \quad (7.2.3)$$

For the solution of Eq. (7.2.3) we put $x_0(\tau) = R_0 \cos\Omega\tau$, where Ω is the dimensionless frequency of the self-excited vibration. After rescaling the parameters with respect to ε as usual, substituting the expression for $x_0(\tau)$ into Eq. (7.2.3), and using averaging or the harmonic balance method (Chapter 9), we obtain the following equations:

$$(q^2 - \Omega^2)R_0 = 0,$$
$$\left[\kappa - \beta V^2\left(1 - \tfrac{1}{4}\gamma R_0^2\right)\right]\Omega R_0 = 0. \quad (7.2.4)$$

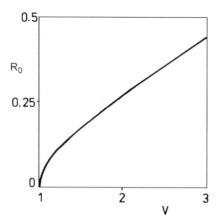

Figure 7.2: Vibration amplitude curve R_0 corresponding to the semitrivial solution as a function of the flow velocity V according to Eqs. (7.2.5). The following values have been used: $\kappa = \beta = 0.02$, $\gamma = 16$.

Assuming that $R_0 \neq 0$, we have

$$\Omega = q,$$
$$R_0^2 = \frac{4}{\gamma}\left(1 - \frac{\kappa}{\beta V^2}\right). \qquad (7.2.5)$$

The requirement of existence of a solution for the second equation of system (7.2.5) implies that $V \geq V_c$, where $V_c = \sqrt{\kappa/\beta}$ is the critical velocity.

As an example, the following parameter values are considered: $\kappa = \beta = 0.02$, $\gamma = 16$, and $\kappa_0 = 0.10$. Figure 7.2 shows the amplitude of the semitrivial solution $R_0(V)$ as a function of the flow velocity V.

We can determine the stability of the semitrivial solution by analysing the differential equations of the disturbed motion that we obtain by inserting into Eqs. (7.2.2) the perturbative solutions

$$x = x_0(\tau) + u,$$
$$\varphi = \varphi_0(\tau) + \psi. \qquad (7.2.6)$$

Thus, for the disturbances u and ψ, we have, after linearisation,

$$u'' + \kappa u' + q^2 u - \beta V^2(1 - \gamma x_0^2)u' + 2\gamma\beta V^2 x_0 x_0' u = 0,$$
$$\psi'' + \kappa_0 \psi' + (1 + x_0'')\psi = 0. \qquad (7.2.7)$$

This is a system of uncoupled Mathieu equations. Approximating the solution on the boundary of the main instability domain by the expressions

$$u(\tau) = a \cos \Omega\tau + b \sin \Omega\tau,$$
$$\psi(\tau) = c \cos \tfrac{1}{2}\Omega\tau + d \sin \tfrac{1}{2}\Omega\tau, \qquad (7.2.8)$$

from the first equation of Eqs. (7.2.7), we obtain the following relation:

$$R^2 = \frac{4}{3\gamma}\left(1 - \frac{\kappa}{\beta V^2}\right), \qquad (7.2.9)$$

and from the second equation of Eqs. (7.2.7) the result is

$$R = \frac{2}{\Omega^2}\left[\left(1 - \tfrac{1}{4}\Omega^2\right)^2 + \tfrac{1}{4}\kappa_0^2\Omega^2\right]^{1/2}. \qquad (7.2.10)$$

Here $R = (a^2 + b^2)^{1/2}$ yields the stability boundary curve. Because the instability domain given by Eq. (7.2.9) lies outside the curve $R_0(V)$ representing the semitrivial solution, relation (7.2.10) is decisive for the stability. From Eqs. (7.2.5) it follows that the stability boundary curve $R = R(\Omega) = R(q)$ can be easily determined from Eq. (7.2.10).

7.3 Nontrivial Solution of the Critical Velocity Model

We look for an approximation of a nontrivial solution of the equations of motion in the form

$$x(\tau) = A \cos \Omega\tau + B \sin \Omega\tau,$$
$$\varphi(\tau) = R_2 \cos \tfrac{1}{2}\Omega\tau. \qquad (7.3.1)$$

Inserting Eqs. (7.3.1) into Eqs. (7.2.2) and using averaging or the harmonic balance method, we obtain the following relations:

$$(q^2 - \Omega^2)A + \left[\kappa - \beta V^2\left(1 - \tfrac{1}{4}\gamma R_1^2\right)\right]\Omega B = \tfrac{1}{4}\mu\Omega^2 R_2^2,$$
$$-\left[\kappa - \beta V^2\left(1 - \tfrac{1}{4}\gamma R_1^2\right)\right]\Omega A + (q^2 - \Omega^2)B = 0,$$
$$\left(1 - \tfrac{1}{4}\Omega^2 - \tfrac{1}{2}\Omega^2 A\right)R_2 = 0,$$
$$-\tfrac{1}{2}\Omega R_2(\kappa_0 + \Omega B) = 0, \qquad (7.3.2)$$

where $R_1 = (A^2 + B^2)^{1/2}$. From the last two equations of Eqs. (7.3.2) we find

$$A = \frac{2}{\Omega^2}\left(1 - \tfrac{1}{4}\Omega^2\right),$$

$$B = -\frac{\kappa_0}{\Omega}, \qquad (7.3.3)$$

and thus

$$R_1 = \frac{2}{\Omega}\left[\left(1 - \tfrac{1}{4}\Omega^2\right)^2 + \tfrac{1}{4}\kappa_0\Omega^2\right]^{1/2}. \qquad (7.3.4)$$

The second equation of Eqs. (7.3.2) yields the relation

$$V^2 = \frac{1}{\beta}\left(1 - \tfrac{1}{4}\gamma R_1^2\right)^{-1}\left[\kappa - \frac{(q^2 - \Omega^2)B}{\Omega A}\right]. \qquad (7.3.5)$$

Multiplying the first equation of Eqs. (7.3.2) by A and the second one by B and adding the results, we obtain the following relation:

$$R_2^2 = \frac{4}{\mu\Omega^2 A}(q^2 - \Omega^2)R_1^2. \qquad (7.3.6)$$

An alternative relation between the amplitudes R_2 and R_1 can be derived when the first equation of Eqs. (7.3.2) is multiplied by B and the second one by A and the final results are added:

$$R_2^2 = \frac{4}{\mu\kappa_0}\left[\beta V^2\left(1 - \tfrac{1}{4}\gamma R_1^2\right) - \kappa\right]R_1^2. \qquad (7.3.7)$$

By using relations (7.3.3), (7.3.4), and (7.3.5) we can calculate expressions for $R_1(\Omega)$ and $V(\Omega)$ for chosen values of q. From Eq. (7.3.6) or (7.3.7) $R_2(\Omega)$ is also derived. For certain values of the velocity V, the corresponding value of Ω can be found and then the values of R_1 and R_2. In this way the expressions for $R_1(V)$, $R_2(V)$, and $\Omega(V)$ can be determined. These expressions for parameter values mentioned in Section 7.3 are represented in Figure 7.3 for the values $q = 1.8$ and $q = 2.2$.

To obtain more insight into the phenomena, Figure 7.3 shows in axonometric representation the stable part of $R_0(V)$ and $R_1(V)$ curves for different values of q. A similar diagram for $R_2(V)$ is given in the same figure. We can see that for $q \approx 2$ the amplitude R_1 is substantially smaller than R_0, from which we conclude that the additional pendulum can substantially diminish the unwelcome self-excited vibration induced by flow.

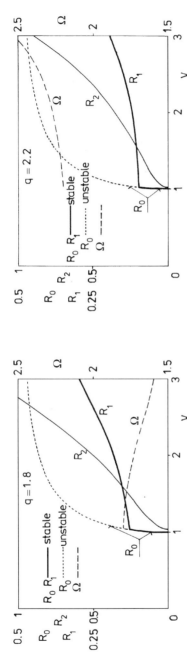

Figure 7.3: Vibration amplitude curves R_0, R_1, and R_2 corresponding to the semitrivial and the nontrivial solutions as a function of the flow velocity V. Stable solutions are marked by heavy solid curves and unstable solutions by dotted curves. The vibration frequency Ω of the nontrivial solution is also plotted. In both diagrams the values of the parameters are $\kappa = \beta = 0.02$, $\gamma = 16$, $\kappa_0 = 0.10$, $\mu = 0.25$. In the left diagram we have taken $q = 1.8$, in the right diagram $q = 2.2$.

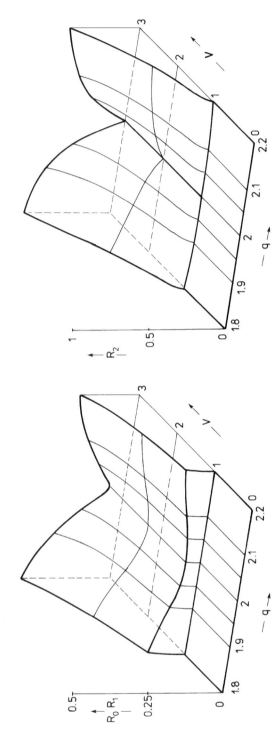

Figure 7.4: In the left diagram the vibration amplitude curves R_0 and R_1 are given, in the right diagram the vibration amplitude curve R_2. These curves correspond to the semitrivial and the nontrivial solutions as a function of the flow velocity V and tuning ratio q. The following values have been used: $\kappa = \beta = 0.02$, $\gamma = 16$, $\kappa_0 = 0.10$, $\mu = 0.25$.

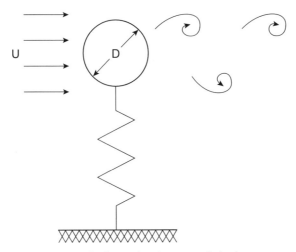

Figure 7.5: The flow-induced system with dry friction.

7.4 The Critical Velocity Model with Dry Friction

A simple mathematical model of a flow-induced system to which a pendulum is attached was investigated in Section 7.3. This section describes the results of a further analysis that has been carried out by considering the additional effect of dry-friction damping acting on the excited subsystem [see Tondl and Nabergoj (1996)]. This new system, after transformation into dimensionless form, is governed by the following differential equations:

$$x'' + \kappa x' + q^2 x - \beta V^2(1 - \gamma x^2)x' + \theta \, \text{sgn}(x') \\ + \mu(\phi'' \sin\phi + \phi'^2 \cos\phi) = 0,$$

$$\phi'' + \kappa_0 \phi' + (1 + x'') \sin\phi = 0. \quad (7.4.1)$$

Here x is the nondimensional displacement of the body excited by the flow and ϕ is the angular deflection of the pendulum, see Figure 7.5. Moreover, κ and κ_0 are the coefficients of linear viscous damping, γ is the coefficient of positive progressive damping, V is the nondimensional relative flow velocity, μ is the ratio of the pendulum mass to the total mass of the system, and θ is the dry-friction coefficient. For $\theta = 0$ the system reduces to system (7.2.2).

System (7.4.1) has the following steady-state solutions:

- the trivial solution $x = x_0 = $ const., $\phi = 0$, where $-\theta < q^2 x_0 < \theta$
- the semitrivial solution $x \neq $ const., $\phi = 0$
- nontrivial solutions $x \neq $ const., $\phi \neq 0$

The stability analysis of the trivial solution for $\theta = 0$ is straightforward and was not considered in Section 7.3. In this particular case, let $V_c = \sqrt{\kappa/\beta}$ be the critical dimensionless velocity; then the trivial solution is unique and stable for $V < V_c$, whereas for $V > V_c$ the trivial solution is unstable. However, for $\theta \neq 0$ the stability investigation of the trivial solution leads to different results and therefore is explicitly considered here.

7.4.1 Stability of the Trivial Solution

After substituting the perturbative solutions

$$x = x_0 + \xi,$$
$$\phi = 0 + \psi, \qquad (7.4.2)$$

into Eqs. (7.4.1) and neglecting higher-order terms, we find the following equations for ξ and ψ:

$$\xi'' + \left[\kappa - \beta V^2 (1 - \gamma x_0^2)\right] \xi' + q^2 \xi + \theta \, \text{sgn}(\xi') = -q^2 x_0,$$
$$\psi'' + \kappa_0 \psi' + \psi = 0. \qquad (7.4.3)$$

Equations (7.4.3) are not coupled. Because all solutions of the equation for ψ converge to zero, the equation for ξ is decisive for the stability. Because of the action of the dry friction, there exists a domain of attraction of initial conditions, leading to a decaying solution converging to x_0.

On the stability boundary, the steady-state solutions can be approximated in the form

$$x_0 = 0,$$
$$\xi = R \cos \Omega \tau, \qquad (7.4.4)$$

where $|x_0| < \theta/q^2$. Thus, after substituting Eqs. (7.4.4) into Eqs. (7.4.3), we obtain the following relations for the oscillation frequency and the vibration amplitude:

$$\Omega = q,$$

$$R = \frac{4\theta}{\pi q [\beta V^2 (1 - \gamma x_0^2) - \kappa]}. \qquad (7.4.5)$$

When $V \to \infty$, the amplitude $R \to 0$. For initial conditions $\xi(0) < R$, $\xi'(0) = 0$, the transient decaying solution, is obtained because positive damping prevails. For $\xi(0) > R$, $\xi'(0) = 0$, the transient solution, diverges. Therefore in the phase plane (ξ, ξ') an unstable limit cycle surrounding the origin exists. This limit cycle is the separatrix forming the boundary of the domain of attraction for the trivial solution.

7.4.2 The Semitrivial Solution

Equations (7.4.1) admit the semitrivial solutions

$$x_0(\tau) = R_0 \cos \Omega \tau,$$

$$\phi_0(\tau) = 0, \qquad (7.4.6)$$

where Ω is the dimensionless frequency of vibration. By inserting the steady-state solution (7.4.6) into Eqs. (7.4.1) and using the harmonic balance method or averaging, we obtain the following equations:

$$(q^2 - \Omega^2) R_0 = 0,$$

$$\left[\kappa - \beta V^2 \left(1 - \tfrac{1}{4}\gamma R_0^2\right)\right]\Omega R_0 + \frac{4}{\pi}\theta = 0, \qquad (7.4.7)$$

from which, for the nontrivial solution, it follows that

$$\Omega = q,$$

$$R_0^3 - \frac{4}{\gamma}\left(1 - \frac{\kappa}{\beta V^2}\right) R_0 + \frac{16}{\pi} \frac{\theta}{\gamma \beta V^2 \Omega} = 0. \qquad (7.4.8)$$

The existence of a solution for the second equation of Eqs. (7.4.8) implies that $V \geq V_t$, where V_t ($> V_c$) is the threshhold velocity. We can show that the semitrivial solution is not stable in the whole range of the flow velocity.

The semitrivial solution is the same as that for the system of 1 degree of freedom obtained when the pendulum is fixed to the body. However, the stability of these systems is different. The higher value of the vibration amplitude belongs to the locally stable solution for the system with 1 degree of freedom, as for the van der Pol system with dry friction. There are two limit cycles: the smaller unstable one is surrounded by a stable limit cycle. The unstable limit cycle forms the boundary of the domain of attraction for the trivial solution. Of course, the trivial solution is stable in both systems in the whole range of flow velocities.

7.4.3 Nontrivial Solutions

The steady-state solutions of the equations of motion can be approximated in the usual way as

$$x(\tau) = A \cos \Omega\tau + B \sin \Omega\tau,$$
$$\phi(\tau) = R_2 \cos \tfrac{1}{2}\Omega\tau. \quad (7.4.9)$$

Inserting Eqs. (7.4.9) into Eqs. (7.4.1) and using the harmonic balance method yields the following relations:

$$(q^2 - \Omega^2)A + \left[\kappa - \beta V^2(1 - \tfrac{1}{4}\gamma R_1^2) + \frac{4}{\pi}\frac{\theta}{\Omega R_1}\right]\Omega B = \tfrac{1}{4}\mu\Omega^2 R_2^2,$$

$$\left[\kappa - \beta V^2(1 - \tfrac{1}{4}\gamma R_1^2) + \frac{4}{\pi}\frac{\theta}{\Omega R_1}\right]\Omega A - (q^2 - \Omega^2)B = 0,$$

$$\left(1 - \tfrac{1}{4}\Omega^2 - \tfrac{1}{2}\Omega^2 A\right)R_2 = 0,$$

$$-\tfrac{1}{2}\Omega R_2(\kappa_0 + \Omega B) = 0, \quad (7.4.10)$$

where $R_1 = (A^2 + B^2)^{1/2}$. From the last two relations of Eqs. (7.4.10) it follows that

$$A = \frac{2}{\Omega^2}(1 - \tfrac{1}{4}\Omega^2),$$
$$B = -\frac{\kappa_0}{\Omega}, \quad (7.4.11)$$

and therefore

$$R_1 = \frac{2}{\Omega^2}\left[(1 - \tfrac{1}{4}\Omega^2)^2 + \tfrac{1}{4}\kappa_0^2\Omega^2\right]^{\tfrac{1}{2}}. \quad (7.4.12)$$

The second equation of Eqs. (7.4.10) yields the relation

$$V^2 = \frac{1}{\beta}(1 - \tfrac{1}{4}\gamma R_1^2)^{-1}\left[\kappa - \frac{(q^2 - \Omega^2)B}{\Omega A} + \frac{4}{\pi}\frac{\theta}{\Omega R_1}\right]. \quad (7.4.13)$$

Multiplying the first equation of Eqs. (7.4.10) by A, the second by B, and adding the results leads to the following relation

$$R_2^2 = \frac{4}{\mu \Omega^2 A}(q^2 - \Omega^2)R_1^2. \quad (7.4.14)$$

An alternative relation between the amplitudes R_1 and R_2 can be derived by multiplying the first equation of Eqs. (7.4.10) by B and the second by A. Subtracting the results yields

$$R_2^2 = \frac{4}{\mu \kappa_0}\left[\beta V^2(1 - \tfrac{1}{4}\gamma R_1^2) - \kappa - \frac{4}{\pi}\frac{\theta}{\Omega R_1}\right]R_1^2. \quad (7.4.15)$$

Using relations (7.4.11)–(7.4.13), we can calculate the functions $R_1(\Omega)$ and $V(\Omega)$ for a given value of q. From Eq. (7.4.14) or (7.4.15) we can also derive $R_2(\Omega)$. For a given value of V, the corresponding value of Ω can be found. In this way, the functions $\Omega(V)$ and $R_1(V)$ can be determined.

7.4.4 Results of the Numerical Investigation

In the calculations, the following parameter values have been used: $\kappa = \beta = 0.02$, $\gamma = 16$, $\kappa_0 = 0.15$, $\mu = 0.25$, and $\theta = 0.005$. The numerical solutions have been calculated for different values of the tuning coefficient, and the results obtained show very similar features. We therefore describe the results for the case $q = 1.9$.

For $V \leq 1.5$, arbitrary initial conditions lead to the trivial solution. When the flow velocity is increased, the self-excited vibration begins at $V = 1.6$, and the vibration records obtained are shown in Figure 7.6. The character of the vibrations is similar to beats but not completely regular. Regular beats are obtained for $V = 1.7$, and the corresponding vibration records are shown in Figure 7.7.

This regular vibration remains unchanged when the flow velocity is increased up to $V = 2.4$. Further increase results in a transient from regular to chaotic vibrations. This is evident from the vibration record for $V = 3.0$, which is shown in Figures 7.8 and 7.9.

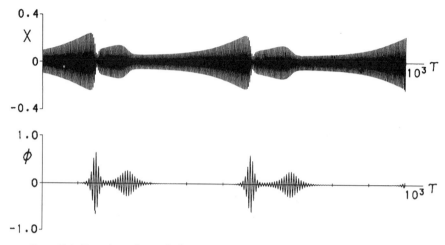

Figure 7.6: Time dependence of vibration amplitudes x and ϕ for flow velocity $V = 1.6$.

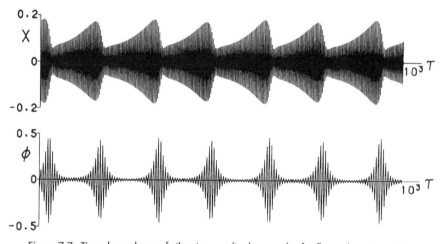

Figure 7.7: Time dependence of vibration amplitudes x and ϕ for flow velocity $V = 1.7$.

The results are summarised in Figure 7.10, in which we show both the analytically computed response curves $R_0(V)$, $R_1(V)$, $R_2(V)$, and the maximal value of the deflections of the body (filled circles) and the pendulum (open circles). For comparison, the diagrams for $\theta = 0$ and $\theta = 0.005$ have been shown. The extreme deflections are larger than the amplitudes obtained analytically; this is due to the beat character of the vibrations. Moreover, for $V = 2.5$, the extreme deflections suddenly increase by an amplitude jump, which is certainly related to the change

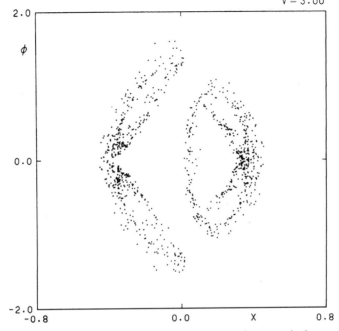

Figure 7.8: The projected Poincaré section $x' = 0$ in the two-dimensional subspace (x, ϕ).

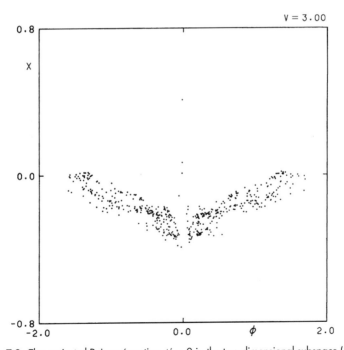

Figure 7.9: The projected Poincaré section $\phi' = 0$ in the two-dimensional subspace (x, ϕ).

Flow-Induced Vibrations

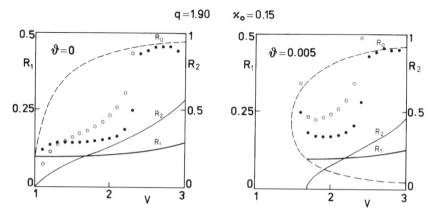

Figure 7.10: Comparison of the numerically computed extreme vibration amplitudes [x] and [ϕ] with the results of analytical predictions. Analytical solutions are marked by solid curves and the results of numerical simulation by circles (filled circles for body and open circles for pendulum).

of the vibration character. We can observe that for higher values of V, the difference between the cases $\theta = 0$ and $\theta = 0.005$ is progressively reduced.

7.4.5 Remarks

From the above-mentioned considerations it follows that the trivial solution is stable in the whole range of the flow velocity. In the situation without damping, there exists a critical velocity V_c, such that when $V > V_c$, another stable solution exists: the nontrivial solution. When dry friction is introduced, the situation is slightly changed. We have found that a stable nontrivial solution exists when $V > V_t$. This threshold velocity V_t is larger than V_c, see, for instance, Figure 7.10, where $V_c = 1.0$ and $V_t = 1.7$.

Because the action of the dry friction, the semitrivial solution is stable in the whole range of flow velocities.

7.5 The Vortex Shedding Model

In this section and in Section 7.6 we perform the elementary analysis of the vortex shedding model: stability of the semitrivial solution and nontrivial solutions. There are many interesting questions left in this model; see also Tondl (1993).

The differential equations of the vortex shedding model with attached pendulum characterised by mass m and pendulum length l are

$$(M+m)\ddot{y} + b\dot{y} + ky + ml(\ddot{\varphi}\sin\varphi + \dot{\varphi}^2\cos\varphi) = \beta_0 u,$$

$$\ddot{u} + \alpha_0 \dot{u} + \omega_s^2[u - \delta\,\text{sgn}(\dot{y})] = 0,$$

$$ml^2\ddot{\varphi} + c\dot{\varphi} + ml(g+\ddot{y})\sin\varphi = 0, \qquad (7.5.1)$$

where c is the damping coefficient of the pendulum motion and g is the acceleration of gravity. Using the time transformation $\tau = \omega_1 t$, we can transform differential equations (7.5.1) into the dimensionless form:

$$x'' + \kappa x' + q^2 x + \mu(\varphi''\sin\varphi + \varphi'^2\cos\varphi) = \beta u,$$

$$u'' + \alpha u' + \omega^2[u - \delta\,\text{sgn}(x')] = 0,$$

$$\varphi'' + \kappa_0 \varphi' + (1+x'')\sin\varphi = 0, \qquad (7.5.2)$$

where $x = y/l$, $\kappa = b/\omega_1(M+m)$, $q = \omega_0/\omega_1$, $\omega_0^2 = k/(M+m)$, $\omega_1^2 = g/l$, $\mu = m/(M+m)$, $\beta = \beta_0/\omega_1 l(M+m)$, $\alpha = \alpha_0/\omega_1$, $\omega = \omega_s/\omega_1$, and $\kappa_0 = c/\omega_1 ml^2$. Equations (7.5.2) admit the semitrivial solutions

$$x_0(\tau) = R_0 \cos\Omega\tau,$$

$$u_0(\tau) = e_0 \cos\Omega\tau + f_0 \sin\Omega\tau,$$

$$\varphi_0(\tau) = 0, \qquad (7.5.3)$$

where Ω is the frequency of vibration. Using again the harmonic balance method or averaging, we obtain the following equations:

$$(q^2 - \Omega^2)R_0 = \beta e_0,$$

$$-\kappa\Omega R_0 = \beta f_0,$$

$$(\omega^2 - \Omega^2)e_0 + \alpha\Omega f_0 = 0,$$

$$-\alpha\Omega e_0 + (\omega^2 - \Omega^2)f_0 = -\frac{4}{\pi}\delta\omega^2. \qquad (7.5.4)$$

The amplitudes e_0 and f_0 in the first two equations of Eqs. (7.5.4) can be substituted into the remaining ones. So the third equation becomes

$$\Omega^4 - (q^2 + \omega^2 + \alpha\kappa)\Omega^2 + q^2\omega^2 = 0, \tag{7.5.5}$$

while the fourth equation gives the relation

$$R_0 = \frac{4}{\pi\Omega}\beta\delta\omega^2[\alpha(q^2 - \Omega^2) + \kappa(\omega^2 - \Omega^2)]^{-1}. \tag{7.5.6}$$

From Eq. (7.5.5) we can determine $\Omega = \Omega(\omega)$ with the other parameters fixed. The amplitude $R_0 = R_0(\omega)$ is obtained with $\Omega(\omega)$, calculated above. Equation (7.5.5) yields two roots, but for $\beta\delta > 0$ only the lowest value of Ω makes sense because we have that $R_0 > 0$. According to Tondl (1992b), we have in this case $\Omega < q$.

The stability investigation of the semitrivial solution leads to a set of three linear, variational differential equations in which the first two, not coupled with the last one, are identical with those obtained when $\varphi = 0$. Tondl (1992b) has shown that, in this particular case, the steady-state solution is stable. So the stability of the semitrivial solution of the extended system can be determined by the analysis of the third variational equation.

After the perturbative solution,

$$\varphi = \varphi_0(\tau) + \psi, \tag{7.5.7}$$

is inserted into Eqs. (7.5.2), the following Mathieu equation is obtained:

$$\psi'' + \kappa_0\psi' + (1 + x_0'')\psi = 0. \tag{7.5.8}$$

As above, we approximate the solutions on the boundary of the main instability domain in the form

$$x(\tau) = R\cos\Omega\tau,$$
$$\psi(\tau) = c\cos\tfrac{1}{2}\Omega\tau + d\sin\tfrac{1}{2}\Omega\tau. \tag{7.5.9}$$

These produce the following equations:

$$\left(1 - \tfrac{1}{4}\Omega^2 - \tfrac{1}{2}\Omega^2 R\right)c + \tfrac{1}{2}\kappa_0\Omega d = 0,$$
$$-\tfrac{1}{2}\kappa_0\Omega c + \left(1 - \tfrac{1}{4}\Omega^2 + \tfrac{1}{2}\Omega^2 R\right)d = 0. \tag{7.5.10}$$

The Vortex Shedding Model

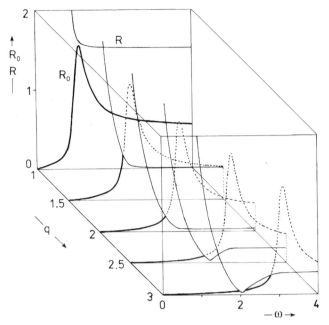

Figure 7.11: Vibration amplitude curve R_0 corresponding to the semitrivial solution (stable solution, heavy solid curves; unstable solution, dotted curves) and stability boundary curve R as a function of the vortex shedding frequency ω and tuning ratio q. The following values have been used: $\alpha = 0.20$, $\beta\delta = 0.02$, $\kappa = \kappa_0 = 0.05$.

The condition for nontrivial solutions c and d leads to the relation

$$R = \frac{2}{\Omega^2}\left[\left(1 - \frac{1}{4}\Omega^2\right)^2 + \frac{1}{4}\kappa_0^2\Omega^2\right]^{1/2}. \quad (7.5.11)$$

Equation (7.5.11) represents the stability boundary curve. As Ω depends on ω, R is also dependent here on ω. The section points of the $R(\omega)$ curves according to Eqs. (7.5.6) and (7.5.11) represent the stability boundaries. To distinguish both curves, the amplitude of the semitrivial solution is marked as R_0 and the stability boundary curve as R in the diagrams. The following parameter values have been taken: $\alpha = 0.2$, $\beta\delta = 0.02$, $\kappa = \kappa_0 = 0.05$. The graphs in Figure 7.11 show $R_0(\omega)$ and $R(\omega)$ for different values of the tuning coefficient q, whereas Figure 7.12 presents the surface of the stable amplitude R_0 as a function of ω and q. We can see that in a relatively broad domain of ω and q, the semitrivial

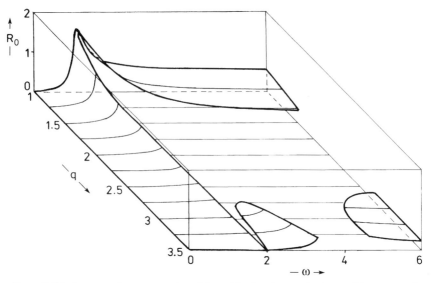

Figure 7.12: Axonometric representation of the stable vibration amplitude R_0 of the semitrivial solution as a function of the vortex shedding frequency ω and tuning ratio q. The following values have been used: $\alpha = 0.20$, $\beta\delta = 0.02$, $\kappa = \kappa_0 = 0.05$.

solution is unstable and the nontrivial solution can be initiated, i.e., the autoparametric resonance can be set on. This instability domain is broadest at q close to 2. In a certain range of q (for $q > 1$) even two instability intervals can exist. The smaller R_0, the narrower the instability domain will be.

7.6 Nontrivial Solution of the Vortex Shedding Model

The steady-state solutions of Eqs. (7.5.2) can be approximated in the form

$$x(\tau) = R_1 \cos \Omega\tau,$$
$$u(\tau) = E \cos \Omega\tau + F \sin \Omega\tau,$$
$$\varphi(\tau) = C \cos \tfrac{1}{2}\Omega\tau + D \sin \tfrac{1}{2}\Omega\tau. \qquad (7.6.1)$$

Using the harmonic balance method or averaging, we obtain the following equations:

$$(q^2 - \Omega^2)R_1 - \beta E = \tfrac{1}{4}\mu\Omega^2(C^2 - D^2),$$

$$-\kappa\Omega R_1 - \beta F = \tfrac{1}{2}\mu\Omega^2 CD,$$

$$(\omega^2 - \Omega^2)E + \alpha\Omega F = 0,$$

$$-\alpha\Omega E + (\omega^2 - \Omega^2)F = -\frac{4}{\pi}\delta\omega^2,$$

$$\left(1 - \tfrac{1}{4}\Omega^2 - \tfrac{1}{2}\Omega^2 R_1\right)C + \tfrac{1}{2}\kappa_0\Omega D = 0,$$

$$-\tfrac{1}{2}\kappa_0\Omega C + \left(1 - \tfrac{1}{4}\Omega^2 - \tfrac{1}{2}\Omega^2 R_1\right)D = 0. \quad (7.6.2)$$

From the last two equations of Eqs. (7.6.2) the condition of C and D, being nontrivial, leads to the relation

$$\left(1 - \tfrac{1}{4}\Omega^2\right)^2 + \tfrac{1}{4}\kappa_0^2\Omega^2 = \tfrac{1}{4}\Omega^4 R_1^2, \quad (7.6.3)$$

from which the function $R_1(\Omega)$ can be determined. From the first two equations of Eqs. (7.6.2) the following relation is obtained:

$$[(q^2 - \Omega^2)R_1 - \beta E]^2 + (\kappa\Omega R_1 - \beta F)^2 = \tfrac{1}{16}\mu^2\Omega^4 R_2^4, \quad (7.6.4)$$

where $R_2 = (C^2 + D^2)^{1/2}$ and E and F are determined from the third and the fourth equations of Eqs. (7.6.2). For very small values of the mass ratio μ the last terms of the first two equations of Eqs. (7.6.2) can be neglected. Then the first four equations of Eqs. (7.6.2) are identical with Eqs. (7.5.4). Because Eq. (7.6.3) is identical with Eq. (7.5.11) we then have the following result: For small values of μ the vibration frequency Ω is close to that calculated from Eq. (7.5.5) and the amplitude $R_1(\omega)$ is well approximated by the stability boundary curve of Eq. (7.5.11).

7.7 Generalisation of the Critical Velocity Model

In this section we study, as a generalisation of the critical velocity model, a 2-degree-of-freedom system involving a self-excited oscillator. Such systems play an important part in mechanical engineering and in studies of vibrations of structures in cross flow but also in other applications; see also Chapters 2 and 3. These systems show new phenomena when compared with other autoparametric systems described in Chapters 2 and 3 and therefore they deserve a separate study. This section is based on the work of Ruijgrok (1995).

First we define a general class of autoparametric systems with self-excitation and derive the corresponding normal-form equations (see Chapter 9 for an introduction to normal forms). The stability of and bifurcations from the semitrivial solution are then studied, leading to various possibilities. As one of the most interesting possibilities, we show the existence of Šilnikov bifurcations in this system and, associated with this bifurcation, chaotic solutions. Although this section is rather technical, it is included to give a mathematical analysis of some of the more complicated dynamics that can occur in autoparametric systems.

7.7.1 The Equations and Their Normal Form

We consider the following system,

$$\ddot{x} + 4x + \varepsilon f(x, \dot{x}) + g(x, y) = 0,$$
$$\ddot{y} + \varepsilon k \dot{y} + (1 + \varepsilon \sigma + \varepsilon a x) V(y) = 0, \quad (7.7.1)$$

where $V(y)$ is a C^∞ function whose Taylor expansion starts with linear terms and ε is a small parameter. The coupling term in the first equation of system (7.7.1) has the form $g(x, y) = g_0(y) + \varepsilon x g_1(y) + \varepsilon^2 x^2 g_2(y) + \cdots$. It is further assumed that the subsystem

$$\ddot{x} + 4x + \delta f(x, \dot{x}) + g(x, 0) = 0 \quad (7.7.2)$$

defines a self-excited oscillator that has a stable π-periodic solution. Note that the critical velocity model studied in Section 7.2 can be put in this form, where the x coordinate describes the self-excited motion of the body in the flow and the y coordinate corresponds to the deflection of the attached (pendulum) subsystem.

As in other autoparametric systems, we expect that the semitrivial solution $(x, \dot{x}, y, \dot{y}) = (x_p(t), \dot{x}_p(t), 0, 0)$ loses stability for certain values of the parameters and that the y mode is then excited. As in the case of the spring–pendulum system (see Chapter 4), it is expected that the resulting oscillations in the y mode will have an amplitude of $\mathcal{O}(\varepsilon^{\frac{1}{2}})$.

System (7.7.1) is studied through its normal form. In essence, the normal form of an equation contains only the linear part and those nonlinear terms that cannot be transformed away. Which nonlinear terms remain depends on only the eigenvalues of the linear part. In practical cases, they can be computed by use of averaging. In this case, these eigenvalues are $\pm i$ and $\pm 2i$. We give the normal form without proof, but details can be found in Ruijgrok (1995). See also Chapter 9 for details on normal forms and averaging.

We make the following assumptions:

1. x and time have been scaled so that subsystem (7.7.2) has a periodic solution of the form $x_p(t) = \cos 2t + \mathcal{O}(\varepsilon)$.
2. $f(x, \dot{x})$ has a Taylor expansion with nontrivial cubic terms. An example is the van der Pol case, where $f(x, \dot{x}) = \dot{x}(1 - x^2)$.

Then, after y is scaled by $\sqrt{\varepsilon}$, Eqs. (7.7.1) are written as a system, and the complex coordinates $z_1 = y - i\dot{y}$ and $z_2 = 2x - i\dot{x}$ are introduced, the normal form of Eqs. (7.7.1), truncated at $\mathcal{O}(\varepsilon)$, becomes

$$\dot{z}_1 = iz_1 + \varepsilon[(-k + i\sigma)z_1 + iA\bar{z}_1 z_2 + iBz_1|z_1|^2],$$
$$\dot{z}_2 = 2iz_2 + \varepsilon[cz_2(1 - |z_2|^2) + ic_0 z_1^2], \quad (7.7.3)$$

with $z_1, z_2 \in \mathbf{C}$ (equations for $\bar{z}_1$ and $\bar{z}_2$ have been omitted). The constants $A, B, k > 0, \sigma, c > 0$, and c_0 are real. It is Eqs. 7.7.3 that we study in this section.

Note the following well-known properties of truncated normal forms of the type of Eqs. (7.7.3): Hyperbolic fixed points, closed orbits, and invariant tori correspond to the same in the original system (7.7.1). Also, the normal form of Eqs. (7.7.3) is invariant under the elements of the one-parameter group of linear transformations $\mathcal{G} \subset Gl(2, \mathbf{C})$, defined by

$$\mathcal{G} = \left\{ g | g = e^{L_0 s}, s \in [0, 2\pi], L_0 = \begin{pmatrix} i & 0 \\ 0 & 2i \end{pmatrix} \right\};$$

see Iooss (1988). In other words, Eqs. (7.7.3) are invariant under $z_1 \to e^{is} z_1, z_2 \to e^{2is} z_2$. This last property is used in Section 7.7.3 to reduce the dimension of the phase space.

7.7.2 Stability of the Semitrivial Solution

Equations (7.7.3) have a π-periodic solution given by $z_1 = 0, z_2 = e^{2it}$. This solution corresponds to the semitrivial solution. To study the stability of this solution, we define $z_2 = e^{2it} + \hat{z}_2$. Equations (7.7.3), linearised near $z_1 = \hat{z}_2 = 0$, become (with the hats dropped)

$$\dot{z}_1 = i z_1 + \varepsilon[(-k + i\sigma)z_1 + i A e^{2it}\bar{z}_1],$$
$$\dot{z}_2 = 2i z_2 - c\varepsilon(z_2 + e^{4it}\bar{z}_2). \tag{7.7.4}$$

Let $z = (z_1, z_1, \bar{z}_1, \bar{z}_2)$. According to Floquet theory (see Verhulst [1996]), the solution of this π-periodic linear equation with initial condition $z = z_0$ can be written as $z = e^{Dt} P(t) z_0$, where D is a constant 2×2 complex matrix and $P(t)$ is a π-periodic 4×4 complex matrix. Let λ_i ($i = 1, \ldots, 4$) be the eigenvalues of $e^{\pi D}$. The λ_i are known as the characteristic multipliers. One of the multipliers will equal 1 (say $\lambda_4 = 1$), because we are linearising near a closed orbit of an autonomous equation. The closed orbit will be stable if $|\lambda_i| < 1$ for $i = 1, 2, 3$. We can find exact solutions of Eqs. (7.7.4) by transforming $z_1 = e^{it} w_1$ and $z_2 = e^{i2t} w_2$, which yields the autonomous equations for w_1 and w_2:

$$\dot{w}_1 = \varepsilon[(-k + i\sigma)w_1 + i A \bar{w}_1],$$
$$\dot{w}_2 = -c\varepsilon(w_2 + \bar{w}_2). \tag{7.7.5}$$

Again, we have omitted the equations for $\bar{w}_1$ and $\bar{w}_2$. Equations (7.7.5) have eigenvalues $\varepsilon(-k \pm \sqrt{\sigma^2 - A^2})$, $-2c\varepsilon$, and 0. The characteristic multipliers of D are therefore

$$\lambda_{1,2} = e^{[i+\varepsilon(-k\pm\sqrt{\sigma^2-A^2})]\pi}, \quad \lambda_3 = e^{(2i-2\varepsilon c)\pi}, \quad \lambda_4 = e^{i2\pi} = 1. \tag{7.7.6}$$

It then follows that the semitrivial solution is stable if

$$A^2 < k^2 + \sigma^2, \tag{7.7.7}$$

a result that is related to the result obtained in Section 7.2 for the critical velocity model.

7.7.3 Reduction of the Normal Form

We can reduce four-dimensional normal-form equations (7.7.3) to a three-dimensional system by first transforming $z_1 = e^{it} w_1$ and $z_2 =$

$e^{2it}w_2$, yielding

$$\dot{w}_1 = \varepsilon[(-k+i\sigma)w_1 + iA\bar{w}_1 w_2 + iBw_1|w_1|^2],$$
$$\dot{w}_2 = \varepsilon\left[cw_2(1-|w_2|^2) + ic_0 w_1^2\right]. \quad (7.7.8)$$

The fact that this transformation leads to an autonomous equation for w_1 and w_2 is a consequence of the invariance of Eqs. (7.7.3) under $z_1 \to e^{is}z_1$, $z_2 \to e^{2is}z_2$. It is easy to see that after a suitable scaling of $w_1, w_2, A, B, k,$ and σ and a time scaling, we can take $c = c_0 = 1$. In polar coordinates, $w_1 = re^{i\phi_1}$ and $w_2 = Re^{i\phi_2}$ and after a time scaling by a factor ε, system (7.7.8) becomes

$$\dot{r} = -kr + ArR\sin(2\phi_1 - \phi_2),$$
$$\dot{\phi}_1 = \sigma + Br^2 + AR\cos(2\phi_1 - \phi_2),$$
$$\dot{R} = R(1-R^2) - r^2\sin(2\phi_1 - \phi_2),$$
$$\dot{\phi}_2 = \frac{r^2}{R}\cos(2\phi_1 - \phi_2). \quad (7.7.9)$$

Because system (7.7.8) is invariant under $w_1 \to e^{is}w_1$, $w_2 \to e^{2is}w_2$, it follows that Eqs. (7.7.9) are invariant under $\phi_1 \to \phi_1 + s$ and $\phi_2 \to \phi_2 + 2s$, which explains why only the combination angle $2\phi_1 - \phi_2$ occurs on the right-hand side of Eqs. (7.7.9). This is typical for resonance problems. Writing $\psi = 2\phi_1 - \phi_2$, we can reduce Eqs. (7.7.9) to the three-dimensional system that is central in this section

$$\dot{r} = -kr + ArR\sin\psi,$$
$$\dot{R} = R(1-R^2) - r^2\sin\psi,$$
$$\dot{\psi} = \left(2AR - \frac{r^2}{R}\right)\cos\psi + 2\sigma + 2Br^2. \quad (7.7.10)$$

We note the following about the interpretation of fixed points of Eqs. (7.7.10).

Let (r_0, R_0, ψ_0) be a fixed point of Eqs. (7.7.10). Equations (7.7.9) then have a solution of the form $r = r_0$, $R = R_0$, $\phi_1 = \alpha t + \phi_0$, and $\phi_2 = 2\alpha t + 2\phi_0 - \psi_0$, where $\alpha = \sigma + AR_0\cos\psi_0 + br_0^2$. Therefore system (7.7.3) has a solution of the form

$$z_1 = r_0 e^{i[(1+\varepsilon\alpha)t+\phi_0]}, \quad z_2 = R_0 e^{2i[(1+\varepsilon\alpha)t+\phi_0-\frac{1}{2}\psi_0]}.$$

These solutions describe a periodic solution with period $[2\pi/(1+\varepsilon\alpha)]$, corresponding to a closed orbit in phase space. As was noted above, if this closed orbit is hyperbolic, then the phase space of the original equations (7.7.1) also contains a hyperbolic closed orbit. The constant ψ_0 is in fact the phase difference between the z_1 and the z_2 modes.

It is, in principle, possible to find all the fixed points of Eqs. (7.7.10) (with $r \neq 0$) analytically. Transforming $u = R\cos\psi$ and $v = R\sin\psi$ yields

$$\dot{r} = -kr + Arv,$$
$$\dot{u} = (1 - R^2)u - 2Auv - (2\sigma + 2Br^2)v,$$
$$\dot{v} = (1 - R^2)v + 2Au^2 + (2\sigma + 2Br^2)u - r^2, \quad (7.7.11)$$

with $R^2 = u^2 + v^2$. Solving

$$-kr + Arv = 0,$$
$$(1 - R^2)u - 2Auv - (2\sigma + 2Br^2)v = 0,$$
$$(1 - R^2)v + 2Au^2 + (2\sigma + 2Br^2)u - r^2 = 0, \quad (7.7.12)$$

yields $v = (k/A)$. Substitution of this expression for v into the second equation of Eqs. (7.7.12) then yields r^2 as a cubic function of u. Finally, substituting this expression into the third equation of Eqs. (7.7.12), we find a quartic equation for u. It therefore follows that there are, at most four fixed points of Eqs. (7.7.10) with $r \neq 0$.

7.7.4 A Fixed Point with One Zero and a Pair of Imaginary Eigenvalues

In theory, the explicit expressions for the fixed points can be used to calculate their stability and determine the various bifurcation scenarios. This method, however, has severe practical difficulties, and we have therefore used numerical software to obtain bifurcation diagrams. A typical diagram corresponding to Eqs. (7.7.10) is shown in Figure 7.13. On curve **a** (curve **b**) the origin, which corresponds to the semitrivial solution, undergoes a supercritical (subcritical) pitchfork bifurcation.

Generalisation of the Critical Velocity Model

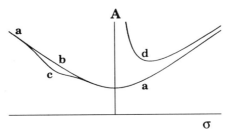

Figure 7.13: Typical bifurcation diagram in the (A, σ) plane, including a secondary Hopf bifurcation.

On curve **c** there is a saddle-node bifurcation of nontrivial fixed points. Finally, inside the hyperbola $A^2 = \sigma^2 + \kappa^2$ there is a curve **d** on which a nontrivial fixed point ($r \neq 0$) undergoes a Hopf bifurcation.

There are points on curve **d** in the (A, σ) plane where the eigenvalues of the linearised vector field are $\pm i\omega$ and 0. The bifurcations associated with this singularity were first studied by Takens (1974) and Guckenheimer (1980). One of the most interesting aspects of this bifurcation is the possibility of a Šilnikov bifurcation. This is a global bifurcation for which it can be shown that near the bifurcation point chaotic dynamics arises; see Šilnikov (1965) and Tresser (1984).

Here we show how such a singularity, which has codimension 2, is unfolded. Because this bifurcation has been studied extensively [see Guckenheimer and Holmes (1983) and Wiggins (1990)], we skip many of the details.

A general formulation for the bifurcation of a fixed point with one zero and a pair of imaginary eigenvalues runs as follows. Consider an equation of the form

$$\dot{w} = L_0 w + F(w), \qquad (7.7.13)$$

where $w = (x, y, z) \in \mathbf{R}^3$, $F(w)$ is a C^∞ function with $F(0) = D_w F(0) = 0$, and

$$L_0 = \begin{pmatrix} 0 & \omega & 0 \\ -\omega & 0 & 0 \\ 0 & 0 & 0 \end{pmatrix}.$$

It is not difficult to show [see Guckenheimer and Holmes (1983)] that, in cylindrical coordinates $x = u \cos\theta$, $y = u \sin\theta$, and $z = z$, the normal form of Eq. (7.7.13) is given by

$$\dot{u} = a_1 uz + a_2 u^3 + a_3 uz^2 + \cdots +,$$
$$\dot{z} = b_1 u^2 + b_2 z^2 + b_3 u^2 z + \cdots +,$$
$$\dot{\theta} = \omega + c_1 z + \cdots + . \qquad (7.7.14)$$

In practical cases, we can find the constants a_1, b_1, and b_2 by transforming to cylindrical coordinates and averaging over θ.

An unfolding of Eqs. (7.7.14) is given by

$$\dot{u} = \mu_1 u + a_1 uz + a_2 u^3 + a_3 uz^2 + \cdots +,$$
$$\dot{z} = \mu_2 + b_1 u^2 + b_2 z^2 + b_3 u^2 z + \cdots +,$$
$$\dot{\theta} = \omega + c_1 z + \cdots + . \qquad (7.7.15)$$

Note that the normal form of Eqs. (7.7.15) has an S^1 symmetry, i.e., it is independent of θ up to any order. Therefore we can study Eqs. (7.7.15) by ignoring the θ component of the vector field and truncating the resulting vector field for (u, z) at a certain order, in effect treating it as the Poincaré map of the three-dimensional normal form. In Section 7.7.7 we will need to consider the effect of the non-S^1-symmetric terms in Eq. (7.7.13). Truncating at order-2 terms and rescaling [see Wiggins (1990)], we can put Eqs. (7.7.15) in the form

$$\dot{u} = \mu_1 u + auz,$$
$$\dot{z} = \mu_2 + bu^2 - z^2, \qquad (7.7.16)$$

with $a = -a_1/b_2$. In Wiggins (1990) the bifurcation diagrams are given for the various possible choices for a and b. It turns out that there are essentially four different types, depending on the signs of a and b.

7.7.5 Normal Form of the Equation

In the case of Eqs. (7.7.10), it can be shown [see Ruijgrok (1995)] that we can choose (A, σ, k) (keeping B unspecified) in such a way that, in normal-form equations (7.7.16), $a = 1$ and $b = \pm 1$ (depending on the value of B). In particular the fact that $a = 1$ makes this a convenient choice, as it will permit some explicit calculations. Using (A, σ)

as unfolding parameters, we then analyse Eqs. (7.7.10), revealing the possibility of Šilnikov bifurcations.

Although the analysis is, strictly speaking, valid for only this particular choice of the parameters, it has a greater generality. Because this bifurcation has codimension 2, its occurrence in the (A, σ) plane is generic. Varying k and B yields another point in the (A, σ) plane where the bifurcation occurs. The difference with the present case is that a different choice for k and B would lead to other values of the normal-form coefficients. However, in the present case we already encounter two bifurcation scenarios, including the most interesting case, which can lead to the Šilnikov bifurcation. The other two bifurcation scenarios show no new phenomena occurring [see Wiggins (1990)]. We believe that it is therefore justified to say that the following analysis is typical for the type of dynamics that can occur in Eqs. (7.7.10).

Applying the procedure described in Section 7.7.4 to Eqs. (7.7.10), we arrive at

$$\dot{u} = \mu_1 u + uz,$$
$$\dot{z} = \mu_2 + bu^2 - z^2, \qquad (7.7.17)$$

where $b = 1$ if $B^2 < 9$ and $b = -1$ when $B^2 > 9$. The parameters μ_1 and μ_2 are differentiable, 1–1 functions of A and σ for fixed k and B.

7.7.6 Bifurcations in the Normal Form

The bifurcations occurring in the normal form of Eqs. (7.7.10) will now be studied.

The Case $b = 1$

First, consider the case in which $B^2 < 9$, corresponding to normal-form equations (7.7.15) with $b = 1$. The bifurcation analysis is straightforward [see Wiggins (1990)] for details and can be summarised in the bifurcation diagram of Figure 7.14(a), which also shows typical phase portraits in the (u, z), $u \geq 0$ half plane.

On the line $\mu_2 = 0$ there occurs a saddle-node bifurcation. On the parabola $\mu_2 = \mu_1^2$, a pitchfork bifurcation occurs. This pitchfork bifurcation is subcritical when $\mu_1 > 0$ and supercritical when $\mu_1 < 0$. It is

Flow-Induced Vibrations

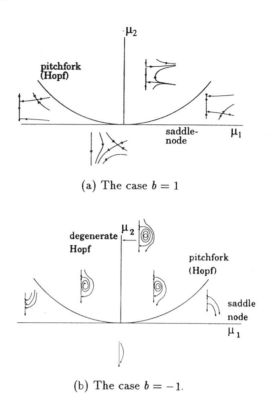

(a) The case $b = 1$

(b) The case $b = -1$.

Figure 7.14: Bifurcation diagram of Eqs. (7.7.17) in the (μ_1, μ_2) plane.

not difficult to see that the fixed points of the planar normal-form equation (7.7.17) have the following interpretation for the three-dimensional normal-form equations (7.7.15): A fixed point with $u = 0$ corresponds to a fixed point of Eqs. (7.7.15) of the same stability type, whereas a fixed point with $u \neq 0$ corresponds to a periodic solution of Eqs. (7.7.15), also of the same stability type. The line $\mu_2 = 0$ therefore corresponds to a saddle-node bifurcation in the full normal form and the parabola $\mu_2 = \mu_1^2$ with a Hopf bifurcation, which is subcritical when $\mu_1 > 0$ and supercritical when $\mu_1 < 0$.

The Case $b = -1$

Now consider the case $b = -1$. The bifurcation diagram for Eqs. (7.7.17), with typical phase portraits, is shown in Figure 7.14(b). Again, there is a saddle-node bifurcation on the line $\mu_2 = 0$, and on the parabola $\mu_2 = \mu_1^2$ a pitchfork bifurcation occurs (in this case subcritical when $\mu_1 < 0$ and

Generalisation of the Critical Velocity Model

supercritical when $\mu_1 > 0$). There is, however, a new phenomenon in this diagram. When $\mu_1 = 0$, the fixed point with $u \neq 0$ undergoes a degenerate Hopf bifurcation, with a typical phase portrait also given in Figure 7.14(b). Note that there exists a homoclinic cycle, consisting of a heteroclinic connection between p_1 and p_2 and the segment of the z axis connecting them and a family of closed orbits surrounding p_3. This very degenerate situation is a result of the fact that Eqs. (7.7.17) are integrable for $\mu_1 = 0$. In fact, putting $\mu_1 = 0$ and transforming $u^2 = x$ (this transformation does not affect the essentials of the dynamics) yields the equations

$$\dot{x} = 2xz,$$
$$\dot{z} = \mu_2 - x - z^2,$$

which can be written as a Hamiltonian system:

$$\dot{x} = \frac{\partial H(x,z)}{\partial z}, \quad \dot{z} = -\frac{\partial H(x,z)}{\partial x},$$

with

$$H(x, z) = -\mu_2 x + \tfrac{1}{2}x^2 + xz^2. \tag{7.7.18}$$

This degeneracy is an effect of the truncation of the planar normal form at terms of degree 2. It is, however, independent of the specific value of the normal-form coefficient a, which in this case equals 1. It is shown in Guckenheimer and Holmes (1983) that the planar normal form with $b = -1$ is integrable when $\mu_1 = 0$ for all values of $a > 0$.

To resolve the degeneracy, we need to consider the normal form, truncated at order 3:

$$\dot{u} = \mu_1 u + uz + (cu^3 + duz^2),$$
$$\dot{z} = \mu_2 - u^2 - z^2 + (eu^2 z + fz^3). \tag{7.7.19}$$

Again in Guckenheimer and Holmes (1983), it is shown that after a (nonlinear) transformation of the coordinates and a time reparameterisation, it can be assumed that $c = d = e = 0$. The calculation of the constant f is a highly nontrivial matter and is not attempted here.

Now we "blow up" the singularity at the origin through the scaling $u = \varepsilon \hat{u}$, $z = \varepsilon \hat{z}$, $\mu_1 = \varepsilon^2 v_1$, and $\mu_2 = \varepsilon^2 v_2$. This scaling is chosen

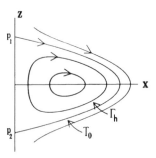

Figure 7.15: Phase portrait of the unperturbed ($\varepsilon = 0$) Hamiltonian system.

because now Eqs. (7.7.19) become a small perturbation of a Hamiltonian system (with the hats dropped):

$$\dot{u} = uz + \varepsilon v_1 u,$$
$$\dot{z} = v_2 - u^2 - z^2 + \varepsilon f z^3. \qquad (7.7.20)$$

Again transforming $x = u^2$, putting $v_2 = 1$ (assuming that $\mu_2 > 0$, which is the case we are interested in, this can be done without loss of generality, as the variation in the original parameter μ_2 is obtained as ε varies), and scaling $v_1 = \frac{1}{2}\hat{v}_1$, we obtain (with the hats dropped)

$$\dot{x} = 2xz + \varepsilon v_1 x,$$
$$\dot{z} = 1 - x - z^2 + \varepsilon f z^3. \qquad (7.7.21)$$

The phase portrait of Eqs. (7.7.21) when $\varepsilon = 0$ consists of the level curves of the Hamiltonian $H(x, z) = -x + \frac{1}{2}x^2 + xz^2$ (see Figure 7.15). The homoclinic cycle T_0 connecting $p_1 = (0, 1)$ and $p_2 = (0, -1)$ is contained in the level curve $H = 0$. The fixed point $(1, 0)$ coincides with the level curve $H = -\frac{1}{2}$, and the closed orbits Γ_h are level curves of $H = h$, with $-\frac{1}{2} < h < 0$.

When $\varepsilon > 0$, the system is no longer integrable, and we expect that almost all closed orbits will break up. A more complete analysis can be obtained with Melnikov's method [see Wiggins (1990)]. The method can be applied to find the values of v_1 for which system (7.7.21) contains a homoclinic cycle, when $\varepsilon > 0$. More generally, by using Melnikov's method we can find the values $v_1(h)$ for which the closed orbit Γ_h survives.

Theorem 1 *Consider a planar equation*

$$\dot{\xi} = f(\xi) + \varepsilon g(\xi, v), \qquad (7.7.22)$$

with $\xi \in \mathbf{R}^2$ and $v \in \mathbf{R}$, which is Hamiltonian when $\varepsilon = 0$. Define

$$M(v) = \int_{\mathrm{int}\,\Gamma} \mathrm{trace}\, Dg(\xi, v)\, d\xi, \qquad (7.7.23)$$

where Γ is a level curve of the unperturbed system (either a closed orbit or a homoclinic cycle). If v_0 is a simple zero of $M(v)$, then for $\varepsilon > 0$ sufficiently small this orbit persists for a value of v near v_0.

Proof [see Guckenheimer and Holmes (1983)].

In the case of Eqs. (7.7.21), the Melnikov function for the homoclinic cycle contained within $H(x, z) = 0$ is given by

$$M(v_1) = \int_{\mathrm{int}\,\Gamma_0} (v_1 + 3fz^2)\, dz\, dx, \qquad (7.7.24)$$

where Γ_0 is defined by $z^2 = 1 - \frac{1}{2}x$. A straightforward integration shows that

$$M(v_1) = \frac{8}{3} v_1 + \frac{8}{5} f, \qquad (7.7.25)$$

which has a simple zero for $v_1 = -\frac{3}{5} f$. In terms of the original parameters, this implies that for $\varepsilon > 0$ sufficiently small, Eqs. (7.7.19) undergo a homoclinic bifurcation on a curve $\mathcal{C}$ in the (μ_1, μ_2) plane, which is of $\mathcal{O}(\varepsilon)$ close to the line $\mu_1 = -\frac{3}{5} f \mu_2$. The stability of the homoclinic cycle is determined by the quantity

$$g = \log \left(\prod_{i=1}^{2} \frac{l_i^1}{l_i^2} \right),$$

where $-l_i^{(1)} < 0 < l_i^{(2)}$, $i = 1, 2$ are the eigenvalues of the saddle points that are connected. A short calculation shows that $g = -\frac{12}{5} \varepsilon f + \mathcal{O}(\varepsilon^2)$, which implies that the homoclinic cycle is stable if $f > 0$ and unstable if $f < 0$.

The Melnikov function for a closed orbit $H = h$ can, after some manipulation, be written as

$$M(v_1) = v_1 \int_\alpha^1 (1-q)^{-\frac{1}{2}} (q^2 - \alpha^2)^{\frac{1}{2}} \, dq$$

$$+ 3f \int_\alpha^1 (1-q)^{\frac{1}{2}} (q^2 - \alpha^2)^{\frac{1}{2}} \, dq,$$

with $\alpha^2 = -2h$. This expression can be evaluated in terms of complete elliptic integrals. However, the main interest is in the zeros of $M(v_1)$, which will be denoted by $v_1(h)$. It has been shown [see Zholondek (1984)] that $v_1(h)$ is a monotone function of h for all values of the normal-form coefficient a, in this case decreasing when $f > 0$ and increasing when $f < 0$, as $v_1(-\frac{1}{2}) = 0$ and $v_1(0) = -\frac{3}{5}f$. This means that for values $-\frac{1}{2} < h < 0$ the Melnikov function has only one zero and therefore only one closed orbit persists.

The complete bifurcation diagram for planar normal-form equations (7.7.19) can now be sketched (see Figure 7.16). When $f > 0$ ($f < 0$), the Hopf bifurcation on the line $\mu_1 = 0$ is subcritical (supercritical) and in the wedge in the (μ_1, μ_2) plane between the line $\mu_1 = -\frac{3}{5} f \mu_2$ and the line $\mu_1 = 0$ a unique repelling (attracting) closed orbit exists. It should be emphasised that this bifurcation diagram is valid for all values of the normal-form coefficient $a > 0$ (with possibly a reflection in the $\mu_1 = 0$ axis).

These results have the following interpretation for the three-dimensional normal-form equations (7.7.15). Closed orbits in the (u, z) plane correspond to invariant tori in (u, z, θ) space. The flow on such a torus is characterised by a fast frequency, corresponding to ω, and a slow frequency, corresponding to the frequency of the closed orbit in the (u, z) plane, which is of $\mathcal{O}(\sqrt{\mu_1})$. The homoclinic cycle corresponds to an invariant sphere filled with heteroclinic orbits connecting the saddle points plus the segment of the z axis connecting the two saddle points. The invariant sphere is simultaneously the two-dimensional unstable manifold $W_u(p_1)$ of point p_1 and the two-dimensional stable manifold $W_s(p_2)$ of point p_2. The coincidence of these manifolds is clearly not generic. In Subsection 7.7.7 we consider what happens in the original, non-S^1-symmetric system (7.7.10).

Generalisation of the Critical Velocity Model

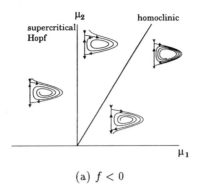

(a) $f < 0$

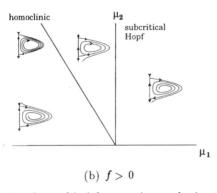

(b) $f > 0$

Figure 7.16: Completion of the bifurcation diagram for the case $b = -1$.

7.7.7 The Possibility of Šilnikov Bifurcations

Consider the three-dimensional normal-form equations (7.7.15) when a small non-S^1-symmetric term is added. Generically, it is expected that the one-dimensional stable manifold $W_s(p_1)$ and the one-dimensional unstable manifold $W_u(p_2)$ will then not intersect. Similarily, it is expected that $W_u(p_1)$ and $W_s(p_2)$ are then transversal and either intersect along heteroclinic orbits or not at all; see Figures 7.17 and 7.18.

In both these cases it may happen that the branch of $W_s(p_1)$ inside the sphere falls into $W_u(p_1)$ or, similarily, the branch of $W_u(p_2)$ inside the sphere falls into $W_s(p_2)$. This then produces a homoclinic orbit (see Figure 7.18). The occurrence of such a homoclinic orbit is referred to as a Šilnikov bifurcation. The possibility of Šilnikov bifurcations in Eqs. (7.7.10) follows from the following result, which has been proved in Broer and Vegter (1984).

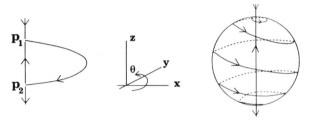

Figure 7.17: The unperturbed case.

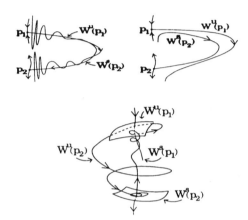

Figure 7.18: The perturbed case. Breakup of the homoclinic cycle and a homoclinic orbit.

Proposition 1 *Let $(\mu_1, \mu_2) \in C$; then there exists a perturbation of normal-form equations (7.7.15) that is flat in r and z [i.e., a perturbation that is asymptotically smaller than any algebraic function of r and z as $(r, z) \to (0, 0)$] such that there is a sequence of points $(\mu_1, \mu_2)_i \in C$ where the perturbed vector field undergoes a Šilnikov bifurcation. Furthermore, given this perturbation, there exists a curve C' that is close to C and is such that the perturbed vector field undergoes a Šilnikov bifurcation when this curve is crossed transversally; see Figure 7.19.*

Normal-form reduction consists of simplifying the Taylor series of a vector field, ignoring the exponentially small flat terms. Therefore the vector field of Eqs. (7.7.10) is equivalent to that of its normal-form equations (7.7.15) up to only flat terms. Consequently, it follows from

Generalisation of the Critical Velocity Model

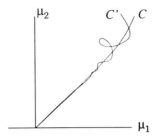

Figure 7.19: Curves $\mathcal{C}$ and $\mathcal{C}'$ in the (μ_1, μ_2) plane.

the proposition that Eqs. (7.7.10) can undergo Šilnikov bifurcations for parameter values such that (μ_1, μ_2) is near $\mathcal{C}$.

The question remains of how generic the perturbation is that leads to a Šilnikov bifurcation. This is a very subtle matter, which is not pursued here as all the details can be found in Broer and Vegter (1984). Suffice it to say that the set of perturbations leading to the Šilnikov bifurcation is C^∞ dense in the space of all perturbations, and therefore the occurrence of the Šilnikov bifurcation in Eqs. (7.7.10) is certainly not very exceptional.

It is well known that the dynamics of vector fields undergoing a Šilnikov bifurcation can be extremely complicated, leading generally to chaotic motion, provided a technical condition on the eigenvalues of the homoclinic point is satisfied. It is an easy task to check that this condition is indeed satisfied for Eqs. (7.7.10). Phase portraits of Eqs. (7.7.10) for values of the parameters given in Subsection 7.7.5 illustrate some of the complex dynamics (see Figure 7.20).

Finally, a remark about the connection between the dynamics of Eqs. (7.7.10), which, after all, are themselves truncated normal-form equations, and the original equations (7.7.1). Solutions of the normal-form equations are close approximations of solutions of the original equations, on a time scale of $1/\varepsilon$. In particular, if the normal form has chaotic solutions (for example, near a Šilnikov bifurcation point), then the original equation will show the same behaviour, at least on a time scale $1/\varepsilon$ [see Sanders and Verhulst (1985)]. Furthermore, hyperbolic fixed points and periodic solutions of Eqs. (7.7.10) correspond to periodic solutions and invariant tori of the normal form of Eqs. (7.7.1) and therefore persist for Eqs. (7.7.1) themselves. Invariant tori of Eqs. (7.7.10) correspond to invariant three tori of the four-dimensional normal form of Eqs. (7.7.1). However, it is known [see Newhouse et al. (1978)] that

127

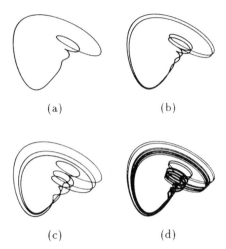

Figure 7.20: Some phase-portraits of Eqs. (7.7.10) projected on the (r, R) plane, showing a sequence of period-doubling bifurcations.

these do not persist for original equations (7.7.1). When Eqs. (7.7.10) undergo a Šilnikov bifurcation, it is known [see Wiggins (1990)] that a countable infinity of horseshoes is created. In Tresser (1984) it is shown that, under generic conditions, for a sufficiently small perturbation of Eqs. (7.7.10) a finite number of these horseshoes persist. Because the original equations (7.7.1) are small perturbations of their normal form (the size of the perturbation can be made arbitrarily small by normalising to a sufficiently high degree), it is therefore conjectured that the original autoparametric problem (7.7.1) possess horseshoes, for most values of the parameters, when Eqs. (7.7.10) undergo a Šilnikov bifurcation.

We hope that the preceding analysis has convinced the reader that it is possible, although not easy, to demonstrate mathematically the occurrence of chaotic solutions in autoparametric systems with self-excitation. However, the application of this analysis to models discussed in this chapter, such as the critical velocity model, has as yet not been attempted.

Chapter 8

Rotor Dynamics

8.1 Introduction

Some rotating machines, e.g., centrifuges, can be modelled by a rigid rotor whose axis of rotation is vertical and is elastically mounted in axial and lateral directions; see Figure 8.1. It is assumed that the axial thrust bearing can be modelled as a joint. The elastic mounting in the axial direction is due to the elasticity of the thrust-bearing support. In some cases the elasticity of the floor on which the machine is situated can influence the elasticity in the axial direction. See Tondl (1988b) for further details of the model. In this chapter we study a basic model for this rigid rotor, which is assumed to be perfectly balanced. In particular, the stability of small vertical oscillations of the upright position are considered. Taking the amplitude of this oscillation as the small parameter and introducing asymptotic expansions around the vertical oscillation leads, to first order, to a system with 2 degrees of freedom, consisting of two coupled Mathieu-like equations. Depending on the frequency of the oscillation and the model's parameters (such as mass, moments of inertia, rotational speed), parametric resonance can occur. Using the method of averaging, we calculate the frequency range for which the motion becomes unstable in Section 8.2.

In Section 8.3 linear damping is added to the model, which leads to remarkable changes in the stability domain.

Rotor Dynamics

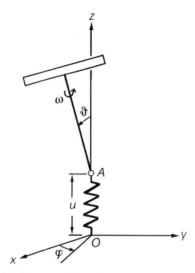

Figure 8.1: The model of the rotor elastically mounted in axial and lateral directions.

Nonlinear damping is added to the model of the rotor in Section 8.4. Numerical bifurcation analysis of the averaged equation shows that the system then exhibits hysteresis and phase locking.

An interesting aspect is that most of the asymptotic expansions for the stable periodic solutions, obtained by averaging in these problems, yield approximations that are valid for all time.

The explanation of the remarkable phenomena found in Section 8.3 requires a subtle mathematical analysis. This is carried out in Section 8.5. The models analysed in this chapter are of course a simplification of real rotor systems. A useful extension would be to include a slight unbalance of the rotor. Furthermore, it would be interesting to consider the effects of autoparametric excitation on other dynamical states of the rotor system, in particular for solutions with precession.

8.2 The Model with Elastic Supports

The following formulation is based on the works of Tondl (1991a, 1992a) and Ruijgrok et al. (1993). Consider a rigid rotor consisting of a heavy disk of mass M that is rotating around an axis (Figure 8.1). The axis of rotation is elastically mounted on a foundation and has a joint in point A; the connections that are holding the rotor in an upright position are also elastic. To describe the position of the rotor we use the axial displacement u in the vertical direction (positive upwards), the angle of the axis of rotation with the z axis and around the z axis. The distance between the centre of gravity B of the rotating disk and point A is R. The moments of inertia are I_1, I_2, and I_3, where, because of the symmetry of the rotor, $I_1 = I_2$.

We derive the equations of motion within a conservative framework by using Lagrange equations, after which we add various kinds of friction. The main purpose is to study the stability of the upright position of the rotor, depending on the system's parameters; so the equations will be linearised around $\theta = 0$, the upright position.

To derive the Lagrange function, the kinetic energy with respect to the reference frame with origin O has to be established. This can be written as

$$T = T_A + \frac{1}{2}M\dot{u}^2 + \dot{u}\sum_i m_i v_i^{(3)}. \quad (8.2.1)$$

The sum, being over all particles in the rotor, gives the coupling term, where T_A is the kinetic energy with respect to A. $v_i^{(3)}$ is the z component of the velocity, with respect to A, of a particle. Because of the symmetry of the rotor, this is equal to the total mass of the rotor times the z component of the velocity of the centre of mass, so that

$$\dot{u}\sum_i m_i v_i^{(3)} = -MR\dot{u}\dot{\theta}\sin\theta. \quad (8.2.2)$$

We wish to study the stability of the upright position of the rotor by considering small oscillations around $\theta = 0$. To have the centre of gravity actually passing through $\theta = 0$, we assume that the angular velocity with respect to the z axis remains constant, say, ω [see Klein and Sommerfeld (1965)]. In that case we can write

$$T_A = \tfrac{1}{2}I_1(\dot{\theta}^2 + \dot{\varphi}^2 \sin^2\theta) + \tfrac{1}{2}I_3[\omega + (\cos\theta - 1)\dot{\varphi}]^2. \quad (8.2.3)$$

The kinetic energy of Eq. (8.2.1) with respect to O then becomes

$$T = \tfrac{1}{2}I_1(\dot{\theta}^2 + \dot{\varphi}^2 \sin^2\theta) + \tfrac{1}{2}I_3[\omega + (\cos\theta - 1)\dot{\varphi}]^2 \\ + \tfrac{1}{2}M\dot{u}^2 - MR\dot{u}\dot{\theta}\sin\theta. \quad (8.2.4)$$

The potential energy is given by

$$V = Mg(R\cos\theta + u) + kR^2\sin^2\theta + k_0 u^2, \quad (8.2.5)$$

where k and k_0 are the coefficients of the lateral stiffness of the mounting and of the vertical stiffness of the axis, respectively.

We look at the projection of the centre of gravity motion on the (x, y) plane, which is given by

$$x = R \sin\theta \cos\varphi,$$
$$y = R \sin\theta \sin\varphi, \tag{8.2.6}$$

which leads to

$$\dot{\theta} = \frac{x\dot{x} + y\dot{y}}{\sqrt{(R^2 - x^2 - y^2)(x^2 + y^2)}}, \quad \dot{\varphi} = \frac{x\dot{y} - y\dot{x}}{x^2 + y^2}.$$

Inserting these expressions into the expressions for the kinetic and the potential energies of Eqs. (8.2.4) and (8.2.5) and retaining the linear and the quadratic terms around $x = y = 0$ give the approximate relations

$$T = \tfrac{1}{2}\frac{I_1}{R^2}(\dot{x}^2 + \dot{y}^2) + \tfrac{1}{2}I_3\left[\omega^2 - \frac{\omega}{R^2}(x\dot{y} - y\dot{x})\right]$$
$$+ \tfrac{1}{2}M\dot{u}^2 - M\frac{\dot{u}}{R}(x\dot{x} + y\dot{y}),$$

$$V = Mg(R + u) + \left(k - \frac{Mg}{2R}\right)(x^2 + y^2) + k_0 u^2. \tag{8.2.7}$$

The Lagrange equations become

$$I_1 \ddot{x} + I_3 \omega \dot{y} + (2kR^2 - MgR)x = MR\ddot{u}x,$$
$$I_1 \ddot{y} - I_3 \omega \dot{x} + (2kR^2 - MgR)y = MR\ddot{u}y,$$
$$\ddot{u} + \frac{2k_0}{M}u = \frac{1}{R}(\dot{x}^2 + x\ddot{x} + \dot{y}^2 + y\ddot{y}) - g. \tag{8.2.8}$$

Dividing the first two equations of Eqs. (8.2.8) by I_1 and using the scaling $t_1 = \Omega t$ with $\Omega^2 = [(2kR^2 - MgR)/I_1]$ finally produces

$$\ddot{x} + 2\alpha\dot{y} + x = \frac{MR}{I_1}\ddot{u}x,$$

$$\ddot{y} - 2\alpha\dot{x} + y = \frac{MR}{I_1}\ddot{u}y,$$

$$\ddot{u} + 4\eta^2 u = \frac{1}{R}(\dot{x}^2 + x\ddot{x} + \dot{y}^2 + y\ddot{y}) - \frac{g}{\Omega^2}. \tag{8.2.9}$$

The Model with Elastic Supports

Here, again, the overdots denote the differentiation with respect to the new time t_1 (later t) and

$$2\alpha = \frac{I_3}{I_1 \Omega} \omega, \quad 4\eta^2 = \frac{2k_0}{\Omega^2 M}.$$

In the following, we consider a rotor system that is externally excited through the support. Thus a special solution to Eqs. (8.2.9) is the semitrivial solution:

$$x_0(t) = 0,$$

$$y_0(t) = 0,$$

$$u_0(t) = a \cos 2\eta t - \frac{g}{4\eta^2 \Omega^2} = a \cos 2\eta t - \frac{Mg}{2k_0}, \quad (8.2.10)$$

corresponding to oscillations in the upright position.

Without loss of generality we have taken the phase of the oscillation as equal to zero. The amplitude a depends on the initial conditions.

We consider the situation in which $a \ll 1$ and study the stability of the semitrivial solution by assuming asymptotic expansions for x, y, and u:

$$x = \varepsilon x_1 + \varepsilon^2 x_2 + \cdots +,$$

$$y = \varepsilon y_1 + \varepsilon^2 y_2 + \cdots +,$$

$$u = -\frac{Mg}{2k_0} + \frac{I_1}{MR} \varepsilon \cos 2\eta t + \varepsilon^2 u_2 + \cdots +, \quad (8.2.11)$$

where

$$\varepsilon = a \frac{MR}{I_1}$$

is a small positive parameter. Inserting expressions (8.2.11) into Eqs. (8.2.9) yields up to first order in ε

$$\ddot{x}_1 + 2\alpha \dot{y}_1 + x_1 = -4\varepsilon \eta^2 x_1 \cos 2\eta t,$$

$$\ddot{y}_1 - 2\alpha \dot{x}_1 + y_1 = -4\varepsilon \eta^2 y_1 \cos 2\eta t. \quad (8.2.12)$$

In the equation for u all terms to order ε vanish. An approximation of $O(\varepsilon^2)$ means the neglect of all nonlinear terms in a neighbourhood of the trivial equilibrium solution $(x, \dot{x}, y, \dot{y}) = (0, 0, 0, 0)$ of Eqs. (8.2.12).

8.3 The Linear System

Replacing (x_1, y_1) with (x, y) in Eqs. (8.2.12), we can write

$$\ddot{x} + 2\alpha \dot{y} + (1 + 4\varepsilon \eta^2 \cos 2\eta t)x = 0,$$
$$\ddot{y} - 2\alpha \dot{x} + (1 + 4\varepsilon \eta^2 \cos 2\eta t)y = 0. \quad (8.3.1)$$

System (8.3.1) constitutes a system of Mathieu-like equations in which we have neglected the effects of damping; see Sections 8.4 and 8.5. The natural frequencies of unperturbed system (8.3.1), $\varepsilon = 0$, are $\omega_1 = \sqrt{\alpha^2 + 1} + \alpha$ and $\omega_2 = \sqrt{\alpha^2 + 1} - \alpha$. It is well known that when the frequency of the autoparametric excitation, 2η, satisfies a resonance condition with the eigenfrequencies of the unperturbed system ($\varepsilon = 0$), then the trivial solution of system (8.3.1), i.e., the semitrivial solution of Eqs. (8.2.9), can become unstable. We shall determine the instability domains for small ε.

By putting $z = x + iy$, we can write system (8.3.1) as

$$\ddot{z} - 2\alpha i \dot{z} + (1 + 4\varepsilon \eta^2 \cos 2\eta t)z = 0. \quad (8.3.2)$$

Introducing the new variable

$$v = e^{-i\alpha t} z \quad (8.3.3)$$

and assuming that $\eta t = \tau$, we obtain

$$v'' + \left(\frac{1 + \alpha^2}{\eta^2} + 4\varepsilon \cos 2\tau \right) v = 0, \quad (8.3.4)$$

where the prime denotes differentiation with respect to τ. By writing down the real and the imaginary parts of this equation, we get two identical Mathieu equations.

We conclude that the trivial solution is stable for ε small enough, providing that $\sqrt{1 + \alpha^2}$ is not close to n, for some $n = 1, 2, 3, \ldots$. The first-order interval of instability, $n = 1$, arises if

$$\sqrt{1 + \alpha^2} \approx \eta. \quad (8.3.5)$$

If approximation (8.3.5) is satisfied, the trivial solution of Eq. (8.3.4) is unstable. Therefore the trivial solution of system (8.3.1) is also unstable.

Note that this instability arises when

$$\omega_1 + \omega_2 = 2\eta,$$

i.e., when the sum of the eigenfrequencies of the unperturbed system equals the autoparametric excitation frequency 2η. This is known as a combination sum–resonance of first order. The domain of instability can be calculated as in Section 9.4; we find for the boundaries

$$\eta_b = \sqrt{1+\alpha^2}\,(1 \pm \varepsilon) + O(\varepsilon^2). \tag{8.3.6}$$

The second-order interval of instability of Eq. (8.3.4), $n = 2$, arises when

$$\sqrt{1+\alpha^2} \approx 2\eta, \tag{8.3.7}$$

i.e., $\omega_1 + \omega_2 \approx \eta$. This is known as a combination sum–resonance of second order. As indicated above, we find the boundaries of the domains of instability:

$$2\eta = \sqrt{1+\alpha^2}\,\left(1 + \tfrac{1}{24}\varepsilon^2\right) + O(\varepsilon^4),$$

$$2\eta = \sqrt{1+\alpha^2}\,\left(1 - \tfrac{5}{24}\varepsilon^2\right) + O(\varepsilon^4). \tag{8.3.8}$$

Higher-order combination resonances can be studied in the same way; the domains of instability in parameter space continue to narrow as n increases. Moreover, it must be kept in mind that the parameter α is proportional to the rotating frequency of the disk and to the ratio of the moments of inertia.

8.4 Stability of the Semitrivial Solution

To examine the stability of the semitrivial solution of Eqs. (8.2.9), we add a small linear damping to system (8.3.1), with positive damping parameter $\mu = 2\varepsilon\kappa$. This leads to

$$\ddot{z} - 2\alpha i\dot{z} + (1 + 4\varepsilon\eta^2 \cos 2\eta t)z + 2\varepsilon\kappa\dot{z} = 0. \tag{8.4.1}$$

Because of the damping term, we can no longer reduce complex equation (8.4.1) to two identical second-order real equations, as we did in Section 8.3.

In the sum–resonance of the first order, $\omega_1 + \omega_2 \approx 2\eta$ and the solution of the unperturbed ($\varepsilon = 0$) equation can be written as

$$z(t) = z_1 e^{i\omega_1 t} + z_2 e^{-i\omega_2 t}, \quad z_1, z_2 \in \mathbf{C}, \qquad (8.4.2)$$

with $\omega_1 = \sqrt{\alpha^2 + 1} + \alpha$ and $\omega_2 = \sqrt{\alpha^2 + 1} - \alpha$. Applying the variation of constants method, we find

$$\dot{z}_1 e^{i\omega_1 t} + \dot{z}_2 e^{-i\omega_2 t} = 0,$$

$$i\omega_1 \dot{z}_1 e^{i\omega_1 t} - i\omega_2 \dot{z}_2 e^{-i\omega_2 t} = -2\varepsilon\kappa(i\omega_1 z_1 e^{i\omega_1 t} - i\omega_2 z_2 e^{-i\omega_2 t})$$
$$+ 4\varepsilon\eta^2 \cos 2\eta t (z_1 e^{i\omega_1 t} + z_2 e^{-i\omega_2 t}),$$

which leads to equations for z_1 and z_2:

$$\dot{z}_1 = \frac{i\varepsilon}{\omega_1 + \omega_2} \{ 2\kappa [i\omega_1 z_1 - i\omega_2 z_2 e^{-i(\omega_1+\omega_2)t}]$$
$$+ 4\eta^2 \cos 2\eta t [z_1 + z_2 e^{-i(\omega_1+\omega_2)t}] \},$$

$$\dot{z}_2 = \frac{-i\varepsilon}{\omega_1 + \omega_2} \{ 2\kappa [i\omega_1 z_1 e^{i(\omega_1+\omega_2)t} - i\omega_2 z_2]$$
$$+ 4\eta^2 \cos 2\eta t [z_1 e^{i(\omega_1+\omega_2)t} + z_2] \}. \qquad (8.4.3)$$

To calculate the instability interval around the value $\eta_0 = \frac{1}{2}(\omega_1 + \omega_2) = \sqrt{\alpha^2 + 1}$, we put

$$\eta = \eta_0 + \varepsilon\sigma, \qquad (8.4.4)$$

where σ is a parameter, independent of ε, that indicates the detuning from exact resonance. In the following we want to obtain the values of σ for which the trivial solutions of Eqs. (8.4.3) become unstable. Inserting Eq. (8.4.4) into Eqs. (8.4.3) yields

$$\dot{z}_1 = \frac{i\varepsilon}{\eta_0} (\kappa(i\omega_1 z_1 - i\omega_2 z_2 e^{-2i\eta_0 t}) + \eta^2 \{ z_1(e^{2i\eta t} + e^{-2i\eta t})$$
$$+ z_2 [e^{2i\varepsilon\sigma t} + e^{-2i(\eta+\eta_0)t}] \}),$$

$$\dot{z}_2 = \frac{-i\varepsilon}{\eta_0} (\kappa(i\omega_1 z_1 e^{2i\eta_0 t} - i\omega_2 z_2) + \eta^2 \{ z_1 [e^{2i(\eta+\eta_0)t} + e^{-2i\varepsilon\sigma t}]$$
$$+ z_2(e^{2i\eta t} + e^{-2i\eta t}) \}). \qquad (8.4.5)$$

Stability of the Semitrivial Solution

After transforming

$$z_1 = v_1 e^{i\varepsilon\sigma t}, \quad z_2 = v_2 e^{-i\varepsilon\sigma t}, \quad (8.4.6)$$

we get equations that are in a suitable form for averaging over t (see Chapter 9):

$$\dot{v}_1 = \frac{\varepsilon}{\eta_0}\{-(\omega_1\kappa + i\sigma\eta_0)v_1 + \kappa\omega_2 v_2 e^{-2i\eta t}$$
$$+ i\eta^2[v_1(e^{2i\eta t} + e^{-2i\eta t}) + v_2(1 + e^{-4i\eta t})]\},$$

$$\dot{v}_2 = \frac{\varepsilon}{\eta_0}\{\kappa\omega_1 v_1 e^{2i\eta t} - (\omega_2\kappa - i\sigma\eta_0)v_2$$
$$- i\eta^2[v_1(1 + e^{4i\eta t}) + v_2(e^{2i\eta t} + e^{-2i\eta t})]\}. \quad (8.4.7)$$

The averaged equations for v_1 and v_2 become

$$\dot{v}_1 = \frac{\varepsilon}{\eta_0}\left[-(\omega_1\kappa + i\sigma\eta_0)v_1 + i\eta_0^2 v_2\right],$$

$$\dot{v}_2 = \frac{\varepsilon}{\eta_0}\left[-i\eta_0^2 v_1 - (\omega_2\kappa - i\sigma\eta_0)v_2\right]. \quad (8.4.8)$$

The stability of the trivial solution of Eqs. (8.4.8) is determined by the real parts of the corresponding eigenvalues.

Let λ' be an eigenvalue of Eqs. (8.4.8). Defining λ by $\lambda' = [(\varepsilon/\eta_0)\lambda]$, the eigenvalue equation for Eqs. (8.4.8) becomes

$$\lambda^2 + 2\eta_0\kappa\lambda + \kappa^2 - 2i\alpha\kappa\sigma\eta_0 + \sigma^2\eta_0^2 - \eta_0^4 = 0, \quad (8.4.9)$$

which has the roots

$$\lambda^{+,-} = -\eta_0\kappa \pm \sqrt{(\alpha\kappa + i\eta_0\sigma)^2 + \eta_0^4}. \quad (8.4.10)$$

For $\kappa = 0$ (no damping), we find that $\lambda^+ = \sqrt{\eta_0^4 - \sigma^2\eta_0^2}$, and the trivial solution is unstable if $|\sigma| < \eta_0 = \sqrt{\alpha^2 + 1}$. This result is consistent with the stability boundary we found in Section 8.3. However, if $\kappa > 0$ we find after some calculations that $\text{Re}(\lambda^+) > 0$ if

$$|\sigma| < \sqrt{\eta_0^4 - \kappa^2}, \quad (8.4.11)$$

and $\eta_0^4 - \kappa^2 > 0$; otherwise there is stability for all values of σ.

In Figure 8.2 we show the boundaries of stable–unstable behaviour in an (ε, η) diagram. If $\kappa = 0$, Eq. (8.3.6) applies; if $\kappa > 0$ we find, by

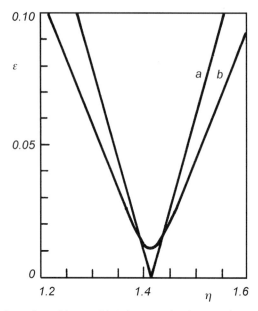

Figure 8.2: Boundaries for stable–unstable behaviour. The diagram shows the instability domains in cases a, $\kappa = 0$ according to Eq. (8.3.6), and b, $\kappa > 0$ according to Eq. (8.4.12) where $\mu = 0.02$. In both cases we have taken $\alpha = 1$.

using Eq. (8.4.4) and inequality (8.4.11), that, to first order,

$$\eta_b = \sqrt{1+\alpha^2}\left(1 \pm \varepsilon \sqrt{1+\alpha^2 - \frac{\kappa^2}{\eta_0^2} + \cdots +}\right),$$

$$= \sqrt{1+\alpha^2}\left[1 \pm \sqrt{(1+\alpha^2)\varepsilon^2 - \left(\frac{\mu}{\eta_0}\right)^2 + \cdots +}\right]. \quad (8.4.12)$$

It follows from Figures 8.2 and 8.3 that the domain of instability actually becomes larger when damping is introduced. This phenomenon can occur only for combination intervals of instability.

The most unusual aspect of the above expression for the instability interval, however, is that there is a discontinuity at $\kappa = 0$. If $\kappa \to 0$, then the boundaries of the instability domain tend to the limits $\eta_b \to \sqrt{1+\alpha^2}(1 \pm \varepsilon\sqrt{1+\alpha^2})$, which differs from the result we found when $\kappa = 0$: $\eta_b = \sqrt{1+\alpha^2}(1 \pm \varepsilon)$. We give a mathematical analysis of this remarkable aspect in Section 8.6.

Nontrivial Solutions: Hysteresis and Phase Locking

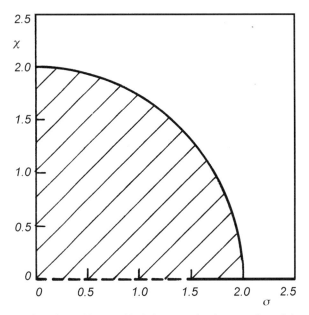

Figure 8.3: Boundary for stable–unstable behaviour; the domain of instability is hatched. The thick and the dashed lines indicate the stability and the instability intervals for $\kappa = 0$, respectively. Again, we have taken $\alpha = 1$.

In mechanical terms, the broadening of the instability domain is caused by the coupling between the 2 degrees of freedom of the rotor in lateral directions that arises in the presence of damping. Such phenomena have been noted earlier in the literature; see Banichuk et al. (1989), Bolotin (1963), or Bratus (1990). The explanation of the discontinuity, however, seems to be new.

8.5 Nontrivial Solutions: Hysteresis and Phase Locking

In this section we consider nontrivial solutions of Eqs. (8.2.9) and examine the related phenomena of hysteresis and phase locking. To obtain a bounded solution, the original equation of motion must be slightly modified, for example, by assuming a progressive damping. Thus we take as the damping function $f(z, \dot{z}) = \kappa \dot{z} + \delta |z|^2 \dot{z}$, with $\kappa, \delta > 0$. After scaling of z by a factor $(k/\delta)^{1/2}$, Eq. (8.4.1) becomes

$$\ddot{z} - 2i\alpha\dot{z} + (1 + 4\varepsilon\eta^2 \cos 2\eta t)z + \varepsilon\kappa\dot{z}(1 + |z|^2) = 0, \quad (8.5.1)$$

and after transforming $w = e^{-i\alpha t} z$ and scaling $\eta t = \tau$ and $\eta = \sqrt{1+\alpha^2}$ $(1+\varepsilon\sigma)$, we find

$$w'' + w + \varepsilon\left[4w\cos 2\tau - 2\sigma w + \kappa\left(\frac{w'}{\sqrt{1+\alpha^2}} + \frac{i\alpha w}{1+\alpha^2}\right)(1+|w|^2)\right] = 0.$$

For $\varepsilon = 0$, the solution of this equation is $w = Ae^{i\tau} + Be^{-i\tau}$, where A and $B \in \mathbf{C}$. Applying the variation of constants to A and B leads to the equations

$$A' = \frac{i\varepsilon}{2}g(A, B, \tau)e^{-i\tau},$$

$$B' = \frac{-i\varepsilon}{2}g(A, B, \tau)e^{i\tau}, \qquad (8.5.2)$$

with

$$g(A, B, \tau) = 4w\cos 2\tau - 2\sigma w + \kappa\left(\frac{w'}{\sqrt{1+\alpha^2}} + \frac{i\alpha w}{1+\alpha^2}\right)(1+|w|^2),$$

where for w and w' we have to substitute $w = Ae^{i\tau} + Be^{-i\tau}$ and $w' = i(Ae^{i\tau} - Be^{-i\tau})$. The right-hand sides of Eqs. (8.5.2) are 2π periodic in time and can therefore be averaged over τ. This leads to the equations

$$A' = \frac{i\varepsilon}{2}\left\{2B - 2\sigma A + \frac{i\kappa}{\eta_0^2}(\eta_0 + \alpha)A + \frac{i\kappa}{\eta_0^2}[(\eta_0 + \alpha)A|A|^2 + 2\alpha A|B|^2]\right\},$$

$$B' = \frac{-i\varepsilon}{2}\left\{2A - 2\sigma B - \frac{i\kappa}{\eta_0^2}(\eta_0 - \alpha)B\right.$$
$$\left. - \frac{i\kappa}{\eta_0^2}[(\eta_0 - \alpha)B|B|^2 - 2\alpha B|A|^2]\right\}, \qquad (8.5.3)$$

where $\eta_0 = \sqrt{1+\alpha^2}$.

Introducing polar coordinates $A = r_1 e^{i\varphi_1}$ and $B = r_2 e^{i\varphi_2}$ yields

$$r_1' = \varepsilon\left[r_2\sin(\varphi_1 - \varphi_2) - \frac{\kappa\omega_1}{2\eta_0^2}r_1(1+r_1^2) - \frac{\kappa\alpha}{\eta_0^2}r_1 r_2^2\right],$$

$$\varphi_1' = \varepsilon\left[\frac{r_2}{r_1}\cos(\varphi_1 - \varphi_2) - \sigma\right],$$

Nontrivial Solutions: Hysteresis and Phase Locking

$$r_2' = \varepsilon\left[r_1 \sin(\varphi_1 - \varphi_2) - \frac{\kappa\omega_2}{2\eta_0^2}r_2(1+r_2^2) + \frac{\kappa\alpha}{\eta_0^2}r_2 r_1^2\right],$$

$$\varphi_2' = \varepsilon\left[\frac{r_1}{r_2}\cos(\varphi_1 - \varphi_2) - \sigma\right]. \qquad (8.5.4)$$

It is easily seen that Eqs. (8.5.4) have a fixed point if $r_1 = r_2$ in this point. This can be the case only when $\alpha = 0$. From the right-hand sides of Eqs. (8.5.4), it is seen that only the phase difference $\psi = \varphi_1 - \varphi_2$ is relevant, so we can study the reduced system:

$$r_1' = \varepsilon\left[r_2 \sin\psi - \frac{\kappa\omega_1}{2\eta_0^2}r_1(1+r_1^2) - \frac{\kappa\alpha}{\eta_0^2}r_1 r_2^2\right],$$

$$r_2' = \varepsilon\left[r_1 \sin\psi - \frac{\kappa\omega_2}{2\eta_0^2}r_2(1+r_2^2) + \frac{\kappa\alpha}{\eta_0^2}r_2 r_1^2\right],$$

$$\psi' = \varepsilon\left[\frac{(r_1^2 + r_2^2)}{r_1 r_2}\cos\psi - 2\sigma\right]. \qquad (8.5.5)$$

Equations (8.5.5) have been investigated with the help of the software package AUTO [see Doedel (1981)]. This program is able to track the fixed points of a system as a parameter is changed (in our case σ), calculate the stability of the fixed point, and, most importantly, detect bifurcations. For system (8.5.5) the results are as follows: the zero solution $r_1 = r_2 = 0$ is unstable if $|\sigma| \leq \sigma_1 = \sqrt{1+\alpha^2 - \frac{1}{4}\kappa^2}$, as we already know from linear analysis; for $0 \leq |\sigma| \leq \sigma_1$ there is also an asymptotically stable fixed point.

For $|\sigma| > \sigma_2$ (see Figures 8.4 and 8.5), only the zero solution is stable and there are no other fixed points. For $\sigma_1 < |\sigma| < \sigma_2$ we have two nonzero solutions, one asymptotically stable and one unstable.

In Figure 8.4 the r_1 and the r_2 components of the fixed points are plotted as functions of σ. The lower half of the "fold," $\sigma_1 < |\sigma| < \sigma_2$, corresponds to the unstable fixed point.

This bifurcation diagram shows that there is hysteresis in system (8.5.5) and therefore also in system (8.5.4). When $|\sigma| < \sigma_1$, system (8.5.5) will tend to the nonzero fixed point. As $|\sigma|$ is increased, this fixed point will remain an attractor until σ_2 is reached, after which this fixed point disappears and the solution will suddenly jump to the zero solution.

Rotor Dynamics

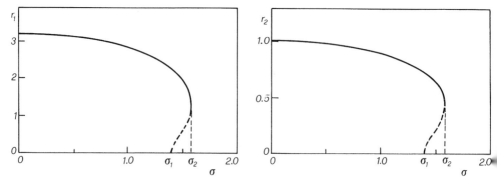

Figure 8.4: Amplitudes r_1 and r_2 of the steady-state periodic solution of Eq. (8.5.1) as functions of the detuning σ (stable solution, heavy solid curves; unstable solution, dashed curves). The parameter values are $\alpha = 1$, $\kappa = 0.02$.

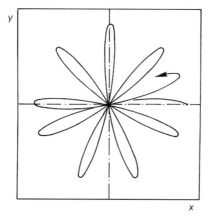

Figure 8.5: Example of a trajectory of the centre of gravity in the (x, y) plane.

System (8.5.4) also exhibits phase locking. When reduced system (8.5.5) tends to a nontrivial fixed point, this implies that the phase difference $\psi = \varphi_1 - \varphi_2$ will converge to a fixed (and in general nonzero) value ψ_0. It then follows from system (8.5.4) that for the asymptotically fixed point we can write

$$\varphi_1(t) = \varepsilon v t, \quad \varphi_2(t) = \varepsilon v t - \psi_0,$$

with $v = (r_1/r_2)\cos\psi_0 - \sigma$. We can now reconstruct the solution of original equation (8.5.1) by inverting the various transformations, and we find that for $|\sigma| \leq \sigma_2$ there exists a stable solution of Eq. (8.5.1) given by

$$z(t) = r_1 e^{i(\omega_1 t + \varepsilon v_1 t)} + r_2 e^{i(-\omega_2 t - \psi_0 + \varepsilon v t)} + \mathcal{O}(\varepsilon) \qquad (8.5.6)$$

on time scale $1/\varepsilon$, with $\omega_1 = \sqrt{\alpha^2 + 1} + \alpha$, $\omega_2 = \sqrt{\alpha^2 + 1} - \alpha$, and r_1, r_2, and ψ_0 [the fixed points of Eqs. (8.5.5)] depending only on σ. Solution (8.5.6) consists of two dominant vibration components, one with forward precession frequency ω_1, the second with backward precession frequency $-\omega_2$. The motion at this resonance is generally nonperiodic, and the trajectory of the centre of gravity in the horizontal plane is not closed; see Figure 8.5 [Tondl (1995)].

For $|\sigma| \geq \sigma_2$, this solution suddenly disappears and only $z(t) = 0$ remains as an asymptotically stable solution.

We conclude that the motion in the axial direction can initiate a whirling motion of the rotor around the axis of rotation. This whirling motion has two components: one consisting of forward, the other of backward precession. The frequencies of these components are different because of the gyroscopic effect of the rotor. When the gyroscopic effect is increasing, the amplitude of the backward precession is growing while the amplitude of the forward precession is decreasing. For more details, see Tondl (1995).

8.6 Parametrically Forced Oscillators in Sum–Resonance

8.6.1 Introduction

In Section 8.3 we came on an unexpected result: When linear damping is added to system (8.3.1), there is a striking discontinuity in the bifurcation diagram, Figure 8.3. Explicitly, linearised equations (8.3.1) are

$$\ddot{x} + 2\alpha \dot{y} + (1 + \varepsilon \cos \omega_0 t)x + 2\mu \dot{x} = 0,$$
$$\ddot{y} - 2\alpha \dot{x} + (1 + \varepsilon \cos \omega_0 t)y + 2\mu \dot{y} = 0, \qquad (8.6.1)$$

where we have put $2\eta = \omega_0$ and replaced $4\varepsilon\eta^2$ with ε.

We assume that ε and μ are small and of the same order, so we scale $\mu = \varepsilon \tilde{\mu}$ and hereafter drop the tilde. The coefficient α is proportional to the speed of rotation of the rotor and is of $\mathcal{O}(1)$ with respect to ε. The eigenfrequencies of Eqs. (8.6.1) for $\varepsilon = 0$ are $\omega_1 = \sqrt{\alpha^2 + 1} + \alpha$ and $\omega_2 = \sqrt{\alpha^2 + 1} - \alpha$, so we find a sum–resonance when ω_0 is near

$\omega_1 + \omega_2 = 2\sqrt{\alpha^2 + 1}$. To allow for detuning we replace ω_1 and ω_2 with $\omega_1 + \delta_1$ and $\omega_2 + \delta_2$; then $\delta_+ = \delta_1 + \delta_2$ detunes the sum–resonance.

For $\mu = 0$, we find that in the normalised equation the origin is unstable if $|\delta_+| \leq 1$. On the other hand, we find that for $\mu > 0$ the origin is unstable if

$$|\delta_+| \leq \omega_0 \sqrt{\frac{1}{4} - \frac{\mu^2}{\omega_0^2}}.$$

Taking the limit $\mu \to 0$, we find that the boundary of the stability interval appears to be $\frac{1}{2}\omega_0 = \sqrt{\alpha^2 + 1}$. This should be contrasted with the inequality $|\delta_+| \leq 1$ we found for $\mu = 0$. We see that the boundary of the instability interval is not a continuous function of μ in $\mu = 0$.

This phenomenon already has been observed and described by Yakubovich and Starzhinskii (1975) and Szemplinska-Stupnicka (1990). Our aim is to present a geometric explanation by using all the parameters as unfolding parameters. It will turn out that four parameters are needed to give a complete description. Fortunately three suffice for visualising the situation.

8.6.2 The General Model

We analyse system (8.6.1) in a slightly more general setting. Consider the following type of differential equation,

$$\dot{z} = A_0 z + \varepsilon f(z, \omega_0 t; \lambda), \quad z \in \mathbf{R}^4, \quad \lambda \in \mathbf{R}^p, \qquad (8.6.2)$$

which describes a system of two parametrically forced coupled oscillators. Here A_0 is a 4×4 matrix with purely imaginary eigenvalues $\pm i\omega_1$ and $\pm i\omega_2$. The vector-valued function f is 2π periodic in $\omega_0 t$ and $f(0, \omega_0 t; \lambda) = 0$ for all t and λ. Equation (8.6.2) can be resonant in many different ways. Because of our rotor problem, we consider the resonance $\omega_1 + \omega_2 = \omega_0$ in which the system exhibits instability. The parameter λ is used to control detuning from resonance and damping.

The first step is to put Eq. (8.6.2) into normal form by normalisation or averaging; see Chapter 9. In Section 8.6.3 we state a normal-form theorem tailored to our situation. We construct the normal form and identify the detuning and the damping parameters. In the normalised equation the time dependence appears in only the high-order terms. But

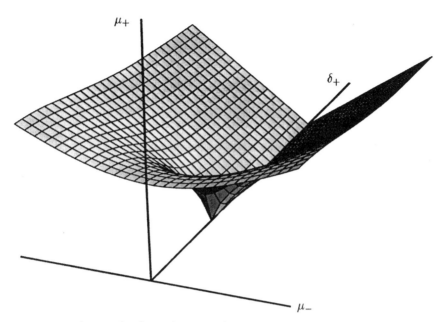

Figure 8.6: The critical surface in (μ_+, μ_-, δ_+) space. $\mu_+ = \mu_1 + \mu_2, \mu_- = \mu_1 - \mu_2, \delta_+ = \delta_1 + \delta_2$. Only the parts $\mu_+ > 0$ and $\delta_+ > 0$ are shown.

the autonomous part of this equation contains enough information to determine the stability regions of the origin.

The second step is to test the linear part $A(\delta, \mu)$ of the normalised equation for structural stability; see Subsection 8.6.4. This family of matrices is parameterised by the detunings δ of ω_1 and ω_2 and the damping parameters μ. We first identify the most degenerate member N of this family and then show that $A(\delta, \mu)$ is its versal unfolding in the sense given by Arnold (1983). Put differently, the family $A(\delta, \mu)$ is structurally stable, whereas $A(\delta, 0)$ is not. Thus we find that the stability diagram actually "lives" in a four-dimensional space. In this space, the stability regions of the origin are separated by a critical surface that is the hypersurface where $A(\delta, \mu)$ has at least one pair of purely imaginary complex-conjugate eigenvalues. This critical surface is diffeomorphic to the Whitney umbrella; see Figure 8.6. For a description of the Whitney (or Whitney–Cayley) umbrella see Arnold (1983); this is a manifold that is used to characterise singularities of maps of two-dimensional manifolds into three-dimensional manifolds. It is the singularity of the Whitney umbrella that causes the discontinuous behaviour of the planar stability diagram. The structural stability argument guarantees that our results are universally valid, i.e., they qualitatively hold for every system in sum–resonance.

8.6.3 The Normalised Equation

Instead of analysing Eq. (8.6.2) we analyse its truncated normal form because it is easier to handle and contains all the essential information. The normal form no longer depends on time and moreover has a symmetry group. We make this more precise by stating a theorem tailored to our situation. Let us first specify the type of equation we consider. The equation is of the following form:

$$\dot{z} = A_0 z + F(z, \mu, t), \quad \dot{\mu} = 0, \quad \dot{t} = 1, \qquad (8.6.3)$$

where F is now 2π periodic in t because time is scaled by a factor ω_0. A_0 is a constant $2n \times 2n$ matrix with purely imaginary eigenvalues. We assume that A_0 is semisimple and hence diagonal. Therefore in complex-conjugate coordinates $z = (x, y) \in \mathbf{C}^n \times \mathbf{C}^n$ with $y = \bar{x}$, we have

$$A_0 = \frac{i}{\omega_0} \operatorname{diag}(\omega_1, \ldots, \omega_n, -\omega_1, \ldots, -\omega_n).$$

Furthermore, we assume that F is C^∞ in all arguments, so that F can be Taylor–Fourier expanded up to any degree l. We find for the jth component ($j = 1, \ldots, n$) that

$$F^{(j)}(z, \mu, t) = \sum_{m=2}^{l} \sum_{\{k, |\gamma| + |\lambda| = m\}} f^{(j)}_{\gamma, \lambda, k} z^\gamma \mu^\lambda e^{ikt} + \tilde{F}^{(j)}_l(z, \mu, t).$$

Here $f^{(j)}_{\gamma, \lambda, k}$ are the Taylor–Fourier coefficients of F and $\tilde{F}^{(j)}_l$ is the remainder term, which is of $\mathcal{O}(|(z, \mu)|^{l+1})$. As usual z^γ means $z_1^{\gamma_1} \cdots z_{2n}^{\gamma_{2n}}$ and $|\gamma|$ means $|\gamma_1| + \cdots + |\gamma_{2n}|$. The notation $\langle \omega, \gamma \rangle$ is shorthand for $\omega_1(\gamma_1 - \gamma_{n+1}) + \cdots + \omega_n(\gamma_n - \gamma_{2n})$. We now state a normal-form theorem for Eq. (8.6.3). For proofs and more details see Arnold (1983) and Broer and Vegter (1992).

Theorem 2 (Normal-Form Theorem) *Consider the differential equation*

$$\dot{z} = A_0 z + F(z, \mu, t),$$

as in Eq. (8.6.3). Then there is a change of coordinates such that in new coordinates we have $\dot{w} = A_0 w + G(w, \mu, t)$ with

$$G^{(j)}(w, \mu, t) = \sum_{m=2}^{l} \sum_{*} g^{(j)}_{\gamma,\lambda,k} w^{\gamma} \mu^{\lambda} e^{ikt} + \tilde{G}^{(j)}_l(w, \mu, t),$$

$$j = 1, \ldots, n,$$

where $\sum_{*}$ is the sum over all k, γ, and λ such that $|\gamma| + |\lambda| = m$ and $R_j(\gamma, k) = \omega_0 k + \langle \omega, \gamma \rangle - \omega_j = 0$. The remainder term $\tilde{G}^{(j)}_l(w, \mu, t)$ is of $\mathcal{O}(|(w, \mu)|^{l+1})$. In the new function G, only resonant terms up to order l appear, that is, terms $w^{\gamma} e^{ikt}$ with $R_j(\gamma, k) = 0$ and $|\gamma| + |\lambda| \le l$. When the coordinate change $z = e^{-tA_0} w$ is used, the truncated equation ($\tilde{G} = 0$) for z becomes time independent:

$$\dot{z}_j = \sum_{m=2}^{l} \sum_{*} g^{(j)}_{\gamma,\lambda,k} z^{\gamma} \mu^{\lambda}, \quad j = 1, \ldots, n.$$

The latter has the symmetry group

$$\mathcal{G} = \{g \mid gz = e^{2\pi i k A_0} z, k \in \mathbf{Z}\},$$

which can be discrete or continuous, depending on the ratios of the ω_i. The equation for z is called the normal form *of Eq. (8.6.3) up to order l.*

We now apply this theorem to Eq. (8.6.2). In this case $n = 2$. Thus the frequencies ω_0, ω_1, and ω_2 determine which terms appear in the normal form. Resonance occurs if the ω satisfy a linear relation with integer coefficients. We assume that there is only one such relation, namely the sum–resonance $\omega_1 + \omega_2 = \omega_0$, where ω_1 and ω_2 are rationally independent. Then the linear part of the normal form is $\dot{z} = A(\delta, \mu) z$ with

$$A(\delta, \mu) = \begin{pmatrix} B(\delta, \mu) & 0 \\ 0 & \overline{B}(\delta, \mu) \end{pmatrix}, \quad (8.6.4)$$

$$B(\delta, \mu) = \begin{pmatrix} i\delta_1 - \mu_1 & \alpha_1 \\ \overline{\alpha}_2 & -i\delta_2 - \mu_2 \end{pmatrix}. \quad (8.6.5)$$

Because $A(\delta, \mu)$ is the complexification of a real matrix, it commutes with complex conjugation. Furthermore, according to the normal-form theorem and with the fact that ω_1 and ω_2 are independent over the integers, the normal form of Eq. (8.6.2) has the continuous symmetry group:

$$\mathcal{G} = \{g \mid gz = \left(e^{is}x_1, e^{is}x_2, e^{-is}y_1, e^{-is}y_2,\right), s \in \mathbf{R}\}.$$

Note that the normal form of the sum–resonance is similar to that of the 1:1 resonance studied by van Gils et al. (1990). Both normal forms are determined by the symmetry group $\mathcal{G}$. Hence the matrices $A(\delta, \mu)$ belong to the set of all $\mathcal{G}$-equivariant matrices that commute with complex conjugation. We call this set **u**. The coordinate transformations that preserve this structure are those that commute with $\mathcal{G}$ and complex conjugation. Thus the group of admissible transformations is given by

$$\left\{ \begin{pmatrix} T & 0 \\ 0 & \overline{T} \end{pmatrix} \,\middle|\, T \in \mathrm{Gl}(2, \mathbf{C}) \right\}.$$

We now discuss the meaning of the parameters $A(\delta, \mu)$. The parameters δ_1 and δ_2 detune the natural frequencies of the oscillators. Note that we do not vary ω_0, but if we set $\delta_+ = \delta_1 + \delta_2$ and $\delta_- = \delta_1 - \delta_2$, then δ_+ detunes the sum–resonance and δ_- detunes the ratio of ω_1 and ω_2. The parameters μ_1 and μ_2 control the damping of the oscillators. For convenience we introduce $\mu_+ = \mu_1 + \mu_2$ and $\mu_- = \mu_1 - \mu_2$. The parameters α_1 and α_2 originate from time-dependent terms in Eq. (8.6.3). They control the strength of the periodic forcing and we assume that they are nonzero. If $\alpha_1 = \alpha_2$, Eq. (8.6.4) is a Hamiltonian matrix for $\mu_1 = \mu_2 = 0$. In this case we find a discontinuity in the planar stability diagram at damping zero (see Section 8.3). As we will see in Section 8.6.4, the discontinuity does not disappear if $\alpha_1 \neq \alpha_2$ but occurs at another value of μ.

8.6.4 Analysis of the Family of Matrices $A(\delta, \mu)$

This subsection is devoted to the analysis of the family $A(\delta, \mu)$. The main results are the following.

Theorem 3 *The family $A(\delta, \mu)$ is equivalent to a versal unfolding $U(\lambda)$ of N in **u** where*

$$N = \begin{pmatrix} N_1 & 0 \\ 0 & N_1 \end{pmatrix}, \quad N_1 = \begin{pmatrix} 0 & 1 \\ 0 & 0 \end{pmatrix}.$$

Corollary 1 *The critical surface of the family $A(\delta, \mu)$ is diffeomorphic to the Whitney umbrella.*

Proof. We prove the theorem by first identifying the most degenerate member N of the family $A(\delta, \mu)$ and then by showing that the family $A(\delta, \mu)$ is equivalent to a versal unfolding $U(\lambda)$ of N. The corollary follows from the fact that the critical surface of $U(\lambda)$ is a Whitney umbrella. In Hoveijn and Ruijgrok (1995) it is shown that the most degenerate member of $A(\delta, \mu)$ is indeed equivalent with N. When Arnold's theory of unfoldings of matrices is used [see Arnold (1983)], a versal unfolding $U(\lambda)$ of N in **u** is given by

$$U(\lambda) = \begin{pmatrix} U_1(\lambda) & 0 \\ 0 & \overline{U}_1(\lambda) \end{pmatrix}, \quad \text{with} \quad U_1(\lambda) = \begin{pmatrix} \lambda_1 & 1 \\ \lambda_2 & \lambda_1 \end{pmatrix}, \quad \lambda_1, \lambda_2 \in \mathbf{C}.$$

For the details of the proof see Hoveijn and Ruijgrok (1995).

The critical surface of $U(\lambda)$ is determined by the characteristic polynomial $p_U(t)$. Although this polynomial has the same roots as those of $A(\delta, \mu)$, it is more convenient to use the real polynomial p_U directly. Let $p_U(t) = t^4 + p_1 t^3 + p_2 t^2 + p_3 t + p_4$; then $U(\lambda)$ has at least one pair of complex-conjugate imaginary eigenvalues if $p_3^2 - p_1 p_2 p_3 + p_1^2 p_4 = 0$. We find this condition by assuming that $p_U(i\beta) = 0$ for $\beta \in \mathbf{R}$ and eliminating β. In $U(\lambda)$, let $\lambda_1 = x + iy$ and $\lambda_2 = u + iv$. Thus the critical surface is given by $x^2(u - x^2) + v^2 = 0$ and the plane $x = 0$, $v = 0$. In (u, v, x) space this surface is diffeomorphic to the Whitney umbrella; see Arnold (1983). For the family of matrices $A(\delta, \mu)$, the critical surface is given by $f_c(\delta, \mu) = \mu_+^4 - \mu_+^2(|\alpha|^2 + \mu_-^2 - \delta_+^2) - \delta_+^2 \mu_-^2 = 0$ if $\alpha_1 = \alpha_2 = \alpha$ [see Hoveijn and Ruijgrok (1995)].

Because $A(\delta, \mu)$ is equivalent to a versal unfolding of N, the family $A(\delta, \mu)$ is structurally stable in the sense given by Arnold (1983). Consequently the family $A(\delta, 0)$ is not structurally stable. These results show that the stability diagram is actually four dimensional. Because the parameter δ_- is relatively unimportant, we can project the four-dimensional stability diagram onto the μ_+, μ_-, δ_+ hyperplane. From now on we consider only this three-dimensional parameter space. Furthermore, for the moment we consider only unfoldings of Hamiltonian systems, that is, $\alpha_1 = \alpha_2 = \alpha$. In this case the stable and the unstable regions of the origin are separated from each other by the critical surface $f_c(\delta, \mu) = 0$. The stable region is given by $f_c(\delta, \mu) > 0$ and $\mu_+ > 0$. The critical surface is shown in Figure 8.6. Because of symmetry we show only the part $\mu_+ > 0$ and $\delta_+ > 0$. This surface consists of three strata: a smooth open part; the half line $\mu_+ = \mu_- = 0$, $\delta_+ > |\alpha|$; and the point $\mu_+ = \mu_- = 0$, $\delta_+ = |\alpha|$. At the half line the surface has a simple self-intersection. The end point of the half line is a singular point.

If $\alpha_1 = \alpha_2 = \alpha$, the line $\mu_+ = \mu_- = 0$ corresponds to the Hamiltonian case. The origin is stable if $\delta_+ > |\alpha|$ for damping zero. However, if we start with nonzero damping and then let damping tend to zero, we end up with a different stability interval. For example, if we set $\mu_+ = a\mu_-$ for $a > 1$, then the boundary of the stability interval is $|\alpha|\{[a^2/(a^2-1)]^{1/2}\}$ for μ_- tending to zero (see Section 8.2). This discontinuity of the boundary of the stability interval is clearly due to the singular point at $\delta_+ = |\alpha|$ of the critical surface.

By allowing $\alpha_1 \neq \alpha_2$, we move the singularity of the critical surface in the μ_-, δ_+ plane. This is another way of perturbing a Hamiltonian system. Qualitatively there is no difference, but the consequences for the planar stability interval are even more drastic because now the origin is unstable in a full neighbourhood of damping zero.

We conclude that the discontinuity in the bifurcation diagram, found in Section 8.3 for the rotor system, is not due to a special choice of coefficients or to the approximation procedure. The discontinuity is a fundamental structural instability in linear gyroscopic systems with at least 2 degrees of freedom and with linear damping.

Chapter 9

Mathematical Methods and Ideas

In this book a number of techniques are being used, in particular, averaging methods, the Poincaré method for determining periodic solutions, and the method of harmonic balance. In this chapter we outline these methods and we indicate their mathematical context; The reader is referred to the literature for more theoretical background. The idea of this chapter is to avoid repetition of calculations and mathematical background in the main text.

We also list some techniques and ideas that we believe are relevant to the study of autoparametric resonance and in general to nonlinear mechanics. In particular we discuss normalisation techniques, bifurcations that arise, and, as a kind of preliminary problem to this book, parametric, nonlinear oscillations.

Notationally we have the following conventions. An overdot denotes the time t derivative, so $\dot{x} = dx/dt$; a prime denotes derivation with respect to nondimensional time τ, so $x' = dx/d\tau$. The parameter ε will always be small:

$$0 \leq \varepsilon \ll 1.$$

9.1 Basic Averaging Results

Averaging is concerned with equations that have been put into the Lagrange standard form:

$$\dot{x} = \varepsilon f(t, x), \quad x(0) = x_0, \tag{9.1.1}$$

where $x \in R^n$, $f(t, x)$ is T periodic in t, and x_0 is the initial value of $x(t)$. We introduce the average

$$f^0(x) = \frac{1}{T} \int_0^T f(t, x) \, dt$$

and consider the associated system

$$\dot{y} = \varepsilon f^0(y), \quad y(0) = x_0. \tag{9.1.2}$$

By solving Eqs. (9.1.2) we find an approximation of $x(t)$, as under rather general conditions we can prove that

$$x(t) - y(t) = \mathcal{O}(\varepsilon) \quad \text{for} \quad 0 \leq \varepsilon t \leq C,$$

where $\mathcal{O}$ is the standard order symbol and C is a positive constant independent of ε. Sometimes this is also expressed verbally as $y(t)$ is an order-ε approximation of $x(t)$ on the time scale $1/\varepsilon$. Note that as ε is a small parameter, this is a long time scale. Proofs can be found in the monograph by Sanders and Verhulst (1985); see also Bogoliubov and Mitropolsky (1961) and Verhulst (1996) for an introduction.

Four remarks should be added.

1. First, $f(t, x)$ might not be T periodic but may have an average in the sense that

$$\lim_{T \to \infty} \frac{1}{T} \int_0^T f(t, x) \, dt = f^0(x)$$

exists. The averaging procedure then also produces an approxima-tion of $x(t)$ on the time scale $1/\varepsilon$; in this case the error may be somewhat larger than ε, for instance $\sqrt{\varepsilon}$ or $\varepsilon \ln \varepsilon$. If $f(t, x)$ is quasi periodic, i.e., f can be expressed as a finite sum of periodic functions with periods independent over the reals, then the error is again of $\mathcal{O}(\varepsilon)$.

2. If Eqs. (9.1.2) contain an equilibrium solution c, so that $f^0(c) = 0$ and c is a hyperbolic fixed point (there are no eigenvalues with real part zero) then there exists a T-periodic solution $\phi(t)$ in an ε neighbourhood of $x_0 = c$. We have $\phi(t) = c + \mathcal{O}(\varepsilon)$ for all time. Moreover, if $y(t) = c$ is an asymptotically stable solution of Eqs. (9.1.2), then any solution $x(t)$ of Eqs. (9.1.1) that starts in the domain of attraction of c is approximated by the solution of the corresponding averaged equation for all time. For a precise formulation and proof see Sanders and Verhulst (1985).

3. The standard-form equation (9.1.1) can be derived from an autonomous system. In this book this takes place, for instance, in the case in which the oscillator is self-excited. If, in such a case, Eqs. (9.1.2) contain an equilibrium solution c, there will always be at least one purely imaginary eigenvalue. If there is only one such purely imaginary eigenvalue, the result mentioned in remark 2 carries over.

4. If necessary, we may calculate higher-order approximations of the solutions of Eqs. (9.1.1). These yield an improvement of the error estimate, but the validity of this result is still on the time scale $1/\varepsilon$.

9.2 Lagrange Standard Forms

Equations like Eqs. (9.1.1) usually arise after a variation-of-constants procedure. Several formulations can be useful; we demonstrate this explicitly for second-order equations that play a prominent part in this book.

Consider the perturbed harmonic oscillator:

$$\ddot{x} + \omega^2 x = \varepsilon F(x, \dot{x}, t). \tag{9.2.1}$$

When $\varepsilon = 0$, the solution can be written as

$$x(t) = r \cos(\omega t + \phi),$$

where r and ϕ are constants determined by the initial conditions.

For Eq. (9.2.1) we introduce, following Lagrange, the transformation $(x, \dot{x}) \to (r, \phi)$ by

$$x(t) = r(t) \cos[\omega t + \phi(t)],$$
$$\dot{x}(t) = -r(t) \omega \sin[\omega t + \phi(t)]. \quad (9.2.2)$$

This is called the amplitude-phase transformation. Requiring that $dx/dt = \dot{x}$ and substituting Eqs. (9.2.2) into Eq. (9.2.1) produces the system in standard form:

$$\dot{r} = -\frac{\varepsilon}{\omega} \sin(\omega t + \phi) F(.,.,t),$$
$$\dot{\phi} = -\frac{\varepsilon}{r\omega} \cos(\omega t + \phi) F(.,.,t). \quad (9.2.3)$$

We have to substitute the appropriate expressions for x and $\dot{x}$ in F. System (9.2.3) is called the Lagrange standard form for averaging.

Another transformation that leads to a standard form and that is particularily useful in the case of parametrically or externally excited systems or in a neighbourhood of $r = 0$ is given by

$$x(t) = a(t) \cos \omega t + b(t) \sin \omega t,$$
$$\dot{x}(t) = -a(t) \omega \sin \omega t + b(t) \omega \cos \omega t. \quad (9.2.4)$$

Using Eq. (9.2.1), we find the standard form

$$\dot{a} = -\frac{\varepsilon}{\omega} F(.,.,t) \sin \omega t,$$
$$\dot{b} = \frac{\varepsilon}{\omega} F(.,.,t) \cos \omega t, \quad (9.2.5)$$

in which we have to substitute the appropriate expressions for x and $\dot{x}$ into $F(.,.,t)$.

9.3 An Example to Illustrate Averaging

Consider the classic problem of a linear, 1-degree-of-freedom oscillator with a small nonlinear force, small damping, and small forcing:

$$\ddot{x} + \varepsilon \kappa \dot{x} + x = \varepsilon f(x) + \varepsilon e \cos t, \quad (9.3.1)$$

with $e, \kappa > 0$. Amplitude-phase transformations (9.2.2) produce

$$\dot{r} = -\varepsilon \sin(t + \phi) \{\kappa r \sin(t + \phi) + f[r \cos(t + \phi)] + e \cos t\},$$

$$\dot{\phi} = -\frac{\varepsilon}{r} \cos(t + \phi) \{\kappa r \sin(t + \phi) + f[r \cos(t + \phi)] + e \cos t\}.$$

Solving this system produces, again by use of transformations (9.2.2), the exact solution of Eq. (9.3.1). We cannot do this, but note that (9.3.2) is in the standard form for averaging.

Indicating by $\tilde{r}$ and $\tilde{\phi}$ the approximations of r and ϕ, we find, after averaging over t, that

$$\frac{d\tilde{r}}{dt} = -\frac{1}{2}\varepsilon(\kappa \tilde{r} + e \sin \tilde{\phi}),$$

$$\frac{d\tilde{\phi}}{dt} = -\frac{1}{2}\varepsilon \frac{e}{\tilde{r}} \cos \tilde{\phi} - \varepsilon \frac{1}{2\pi \tilde{r}} \int_0^{2\pi} \cos(t + \tilde{\phi}) f[\tilde{r} \cos(t + \tilde{\phi})] \, dt,$$

where we used

$$\int_0^{2\pi} \sin(t + \tilde{\phi}) f[\tilde{r} \cos(t + \tilde{\phi})] \, dt = 0.$$

We consider two cases.

Case 1 The function $f(x)$ is even. In this case the averaged phase equation becomes simply

$$\frac{d\tilde{\phi}}{dt} = -\frac{1}{2}\varepsilon \frac{e}{\tilde{r}} \cos \tilde{\phi}.$$

Equilibrium solutions are obtained from

$$\kappa \tilde{r} + e \sin \tilde{\phi} = 0,$$

$$\cos \tilde{\phi} = 0.$$

We find one set of roots with $\tilde{r} > 0$, namely,

$$\tilde{r} = \frac{e}{\kappa}, \quad \tilde{\phi} = \frac{3\pi}{2}.$$

Linearisation of the averaged equations and substitution of $\tilde{r} = e/\kappa$ and $\tilde{\phi} = 3\pi/2$ produce a matrix with negative eigenvalues. We conclude that this equilibrium corresponds with a periodic solution of original equation (9.3.1), which is asymptotically stable. The approximation

$$\tilde{x}(t) = \frac{e}{\kappa} \cos\left(t + \frac{3\pi}{2}\right) \quad (9.3.2)$$

is valid for all time.

Case 2 The function $f(x)$ contains odd terms. In this case the averaged phase equation becomes slightly more complicated. Consider for instance the case of Duffing's equation, where $f(x) = x^3$. We find for the averaged system

$$\frac{d\tilde{r}}{dt} = -\frac{1}{2}\varepsilon(\kappa\tilde{r} + e \sin\tilde{\phi}),$$

$$\frac{d\tilde{\phi}}{dt} = -\frac{1}{2}\varepsilon\left(\frac{e}{\tilde{r}}\cos\tilde{\phi} + \frac{3}{4}\tilde{r}^2\right). \quad (9.3.3)$$

Equilibrium solutions of system (9.3.3) are obtained from

$$\kappa\tilde{r} + e \sin\tilde{\phi} = 0,$$

$$e \cos\tilde{\phi} + \frac{3}{4}\tilde{r}^3 = 0. \quad (9.3.4)$$

Suppose we have found an equilibrium solution $\tilde{r} = r_0$, $\tilde{\phi} = \phi_0$ from Eqs. (9.3.4). The linearisation of system (9.3.3) near (r_0, ϕ_0) is given by the matrix of coefficients,

$$\begin{pmatrix} -\kappa & -e \cos\phi_0 \\ \frac{e}{r_0^2}\cos\phi_0 - \frac{3}{2}r_0 & \frac{e}{r_0}\sin\phi_0 \end{pmatrix}, \quad (9.3.5)$$

where we omitted the factors $\frac{1}{2}\varepsilon$. Using Eqs. (9.3.4) we find

$$\begin{pmatrix} -\kappa & \frac{3}{4}r_0^3 \\ -\frac{9}{4}r_0 & -\kappa \end{pmatrix}, \quad (9.3.6)$$

with eigenvalues $-\kappa \pm \frac{3}{4}\sqrt{3}r_0^2 i$. As above, we have that such an equilibrium solution of system (9.3.3) corresponds to a periodic solution of system (9.2.3), which is asymptotically stable. The corresponding approximation

$$\tilde{x}(t) = r_0 \cos(t + \phi_0) \quad (9.3.7)$$

is valid for all time.

9.4 The Poincaré–Lindstedt Method

The averaging method is useful if one has to study the evolution in time for solutions with general initial conditions. At the same time, as we have seen, we can extract special solutions such as the periodic ones.

In a number of problems our attention is focussed mainly on periodic solutions, and in such a case the Poincaré–Lindstedt method or continuation method is quite efficient. The method applies to nonlinear equations of arbitrary dimension, but we demonstrate its use for a type of problem that arises again and again in this book: equations of Mathieu type.

For a more general treatment the reader should consult Schmidt (1975), Roseau (1966), Hale (1969), or Verhulst (1996); detailed examples and many references to theory and applications can be found in Nayfeh and Mook (1979).

The Poincaré–Lindstedt method is used to approximate periodic solutions of systems of the form

$$\dot{x} = Ax + \varepsilon B(t)x, \qquad (9.4.1)$$

in which $x \in R^n$, A is a constant $n \times n$ matrix, and $B(t)$ is a continuous T-periodic $n \times n$ matrix. Adding small nonlinearities poses no essential obstruction, but in this section we keep the equations linear.

Floquet theory tells us that the solutions of Eq. (9.4.1) can be written as

$$x(t) = \Phi(t)e^{Ct}, \qquad (9.4.2)$$

where $\Phi(t)$ is a T-periodic $n \times n$ matrix and C is a constant $n \times n$ matrix. The determination of C provides us with the stability behaviour of the solutions.

A particular case of Eq. (9.4.1) is Hill's equation:

$$\ddot{x} + b(t)x = 0, \qquad (9.4.3)$$

which is of second order; $b(t)$ is a scalar T-periodic function. A particular case of Eq. (9.4.3) that arises frequently in applications is the Mathieu equation:

$$\ddot{x} + (a + \varepsilon \cos 2t)x = 0, \quad a > 0. \qquad (9.4.4)$$

The following is a typical question: For which values of a and ε in (a, ε) parameter space is the trivial solution $x = \dot{x} = 0$ stable?

Solutions of Eq. (9.4.4) can be written in the Floquet form of Eq. (9.4.2), in which, in this case, $\Phi(t)$ will be π periodic. The eigenvalues λ_1 and λ_2 of C, which are called characteristic exponents, determine the stability of the trivial solution. For the characteristic exponents of Eq. (9.4.1) we have

$$\sum_{i=1}^{n} \lambda_i = \frac{1}{T} \int_0^T Tr[A + \varepsilon B(t)] \, dt; \qquad (9.4.5)$$

see theorem 6.6 in Verhulst (1996). Therefore in the case of Eq. (9.4.4) we have

$$\lambda_1 + \lambda_2 = 0. \qquad (9.4.6)$$

The exponents are functions of ε, $\lambda_1 = \lambda_1(\varepsilon)$, and $\lambda_2 = \lambda_2(\varepsilon)$, and clearly $\lambda_1(0) = i\sqrt{a}$ and $\lambda_2(0) = -i\sqrt{a}$. As $\lambda_1 = -\lambda_2$, the characteristic exponents, which are complex conjugates, are purely imaginary or real. The implication is that if $a \neq k^2$, $k = 1, 2, \ldots$, the characteristic exponents are purely imaginary and $x = 0$ is stable near $\varepsilon = 0$. If $a = k^2$ for some $k \in N$, however, the imaginary part of $\exp(Ct)$ can be absorbed into $\Phi(t)$ and the characteristic exponents may be real.

We assume now that $a = k^2$ for some $k \in N$, or near this value, and we look for periodic solutions of $x(t)$ of Eq. (9.4.4). We put

$$a = k^2 - \varepsilon\beta, \qquad (9.4.7)$$

where β is a constant, and we use transformations (9.2.4) in the form

$$x(t) = y_1(t) \cos kt + y_2(t) \sin kt,$$
$$\dot{x}(t) = -y_1(t)k \sin kt + y_2(t)k \cos kt, \qquad (9.4.8)$$

to obtain from Eq. (9.4.4)

$$\dot{y}_1 = -\frac{\varepsilon}{k}(\beta - \cos 2t)[y_1(t) \cos kt + y_2(t) \sin(kt)] \sin kt,$$

$$\dot{y}_2 = \frac{\varepsilon}{k}(\beta - \cos 2t)[y_1(t)k \cos kt + y_2(t) \sin(kt)] \cos kt. \qquad (9.4.9)$$

It is clear from Eq. (9.4.7) that $a = a(\varepsilon)$, with $a(0) = k^2$; also, from this

The Poincaré–Lindstedt Method

assumption, $\beta = \beta(\varepsilon)$. Poincaré showed that $[x(t), \dot{x}(t)]$ and so $[y_1(t), y_2(t)]$ have a convergent analytic expansion in powers of ε. We therefore write

$$a = k^2 + a_1\varepsilon + a_2\varepsilon^2 + \cdots +, \quad \beta = \beta_0 + \beta_1\varepsilon + \beta_2\varepsilon^2 + \cdots +.$$

For Eq. (9.4.4) to have periodic solutions, Eqs. (9.4.9) must have periodic solutions. We expand with respect to ε and require that at each power of ε the secular (unbounded) terms be eliminated; this leaves the periodic terms.

The case $k=1$
The periodicity conditions from Eqs. (9.4.9) are

$$\int_0^{2\pi} (\beta - \cos 2t)[y_1(t)\cos t + y_2(t)\sin t]\sin t \, dt = 0,$$

$$\int_0^{2\pi} (\beta - \cos 2t)[y_1(t)\cos t + y_2(t)\sin t]\cos t \, dt = 0, \quad (9.4.10)$$

where we still have to expand β, y_1, and y_2 with respect to powers of ε. We find, at lowest order,

$$y_2(0)\left(\beta_0 + \frac{1}{2}\right) = 0,$$

$$y_1(0)\left(\beta_0 - \frac{1}{2}\right) = 0,$$

so that $y_1(0) = 0$, $\beta_0 = -\frac{1}{2}$ or $y_2(0) = 0$, $\beta_0 = \frac{1}{2}$. Periodic solutions exist for $k = 1$ if

$$a = 1 \pm \frac{1}{2}\varepsilon + \mathcal{O}(\varepsilon^2).$$

The case $k=2$
The periodicity conditions from Eqs. (9.4.9) are

$$\int_0^{2\pi} (\beta - \cos 2t)[y_1(t)\cos 2t + y_2(t)\sin 2t]\sin 2t \, dt = 0,$$

$$\int_0^{2\pi} (\beta - \cos 2t)[y_1(t)\cos 2t + y_2(t)\sin 2t]\cos 2t \, dt = 0. \quad (9.4.11)$$

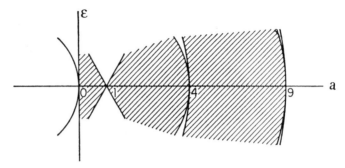

Figure 9.1: Domains of stability for the Mathieu equation (9.4.4) are shaded in (a, ε) parameter space.

At the lowest order we find $\beta_0 = 0$; at the next order we have that periodic solutions exist if $\beta_1 = \frac{1}{48}\varepsilon$ or $\beta_1 = -\frac{5}{48}\varepsilon$. Calculating $\mathcal{O}(\varepsilon^2)$ produces $\beta_2 = 0$. Therefore, periodic solutions exist if

$$a = 1 - \tfrac{1}{48}\varepsilon^2 + \mathcal{O}(\varepsilon^4),$$
$$a = 1 + \tfrac{5}{48}\varepsilon^2 + \mathcal{O}(\varepsilon^4). \qquad (9.4.12)$$

The corresponding stability domains have been depicted in Figure 9.1.

On considering higher values of k, we have to calculate to a higher order of ε. At $k=1$ the boundary curves are intersecting at positive angles at $\varepsilon = 0$; at $k=2$ ($a=4$) they are tangent; the order of tangency increases with k, making instability domains more and more narrow for small values of ε.

9.5 The Mathieu Equation with Viscous Damping

In the applications we are considering in this book, there is always the presence of damping. We consider the effect of its simplest form, small viscous damping. Equation (9.4.4) is extended by the addition of a linear damping term:

$$\ddot{x} + \kappa\dot{x} + (a + \varepsilon\cos 2t)x = 0, \quad a, \kappa > 0. \qquad (9.5.1)$$

We assume that the damping coefficient is small, $\kappa = \varepsilon\kappa_0$, and again we put $a = k^2 - \varepsilon\beta$. Applying transformations (9.4.8), we obtain

$$\dot{y}_1 = \varepsilon \frac{\kappa_0}{k}(-ky_1 \sin kt + ky_2 \cos kt) \sin kt$$
$$\quad - \frac{\varepsilon}{k}(\beta - \cos 2t)[y_1(t)\cos kt + y_2(t)\sin kt]\sin kt,$$

$$\dot{y}_2 = -\varepsilon \frac{\kappa_0}{k}(-ky_1 \sin kt + ky_2 \cos kt)\cos kt$$
$$\quad + \frac{\varepsilon}{k}(\beta - \cos 2t)[y_1(t)k\cos kt + y_2(t)\sin kt]\cos kt. \quad (9.5.2)$$

The case $k = 2$
Imposing the periodicity condition on system (9.5.2) yields, at lowest order,

$$-\tfrac{1}{2}\kappa_0 y_1(0) - y_2(0)\left(\beta_0 + \tfrac{1}{2}\right) = 0,$$
$$-\tfrac{1}{2}\kappa_0 y_2(0) + y_1(0)\left(\beta_0 - \tfrac{1}{2}\right) = 0. \quad (9.5.3)$$

Nontrivial solutions of Eqs. (9.5.3) exist if

$$\kappa_0^2 + \beta_0^2 - \frac{1}{4} = 0.$$

Returning to the original parameters, we find periodic solutions in the case $k = 1$ if

$$a = 1 \pm \sqrt{\frac{1}{4}\varepsilon^2 - \kappa^2}. \quad (9.5.4)$$

Relation (9.5.4) corresponds with the curve of periodic solutions, which in (a, ε) parameter space separates stable and unstable solutions; see Figure 9.2. We observe the following phenomena. If $0 < \kappa < \tfrac{1}{2}\varepsilon$, we have an instability domain that by damping has been lifted from the a axis; also the width has shrunk. If $\kappa > \tfrac{1}{2}\varepsilon$ the instability domain has vanished.

Repeating the calculations for $k \geq 2$, we find no instability domains at all; damping of $\mathcal{O}(\varepsilon)$ stabilises the trivial solution. To find an instability domain we have to decrease the damping; for instance, if $k = 2$, we have to take $\kappa = \varepsilon^2 \kappa_0$. The actual calculations are left to the reader.

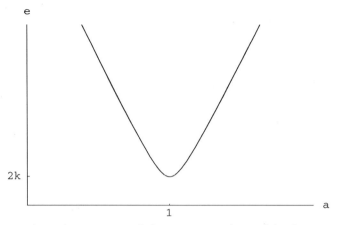

Figure 9.2: In the Mathieu equation with damping (9.5.1), the instability domains are lifted from the a axis.

9.6 The Method of Harmonic Balance

When looking for a periodic solution of a differential equation, we can start with a Fourier expansion of this solution of the form

$$x(t) = \sum_{n=0}^{\infty}(a_n \cos n\Omega t + b_n \sin n\Omega t)$$

or

$$x(t) = \sum_{n=0}^{\infty} R_n \cos(n\Omega t + \phi_n).$$

In some problems we are looking for a solution with a certain period and in such a case Ω is known; if not, Ω will be one of the unknowns to be determined. The idea of harmonic balance is to insert the Fourier expansion into the differential equation and to determine the unknowns (a_n, b_n, Ω) by equating the coefficients of equal harmonics.

Let us say at the outset that the method of harmonic balance produces trustworthy results only if we have enough a priori knowledge of the periodic solution; this requires a lot of experience, especially in the case of more degrees of freedom. But even in this case we still have to clarify in what sense an approximation of the periodic solution has been achieved.

The Method of Harmonic Balance

The reader is also referred to a basic paper by Urabe (1965) and to a useful discussion in Nayfeh and Mook (1979). These authors stress that one has to check the order of magnitude of the coefficients of neglected harmonics. This is a highly nontrivial affair.

We consider two examples.

Example 1
We are looking for periodic solutions of the equation

$$\ddot{x} + x = \varepsilon x^3. \qquad (9.6.1)$$

As the equation is autonomous, we choose the phase $\phi_0 = 0$ in a one-mode expansion:

$$x(t) = R_0 \cos \Omega t.$$

Substitution this mode into the equation (9.6.1) yields

$$-R_0 \Omega^2 \cos \Omega t + R_0 \cos \Omega t = \varepsilon R_0^3 \cos^3 \Omega t$$
$$= \varepsilon R_0^3 \left(\tfrac{3}{4} \cos \Omega t + \tfrac{1}{4} \cos 3\Omega t \right).$$

Equating the coefficients of $\cos \Omega t$, we find

$$-\Omega^2 + 1 = \tfrac{3}{4} \varepsilon R_0^2,$$

or

$$\Omega = 1 - \tfrac{3}{8} \varepsilon R_0^2 + \varepsilon^2 + \cdots +. \qquad (9.6.2)$$

Note that the amplitude R_0 is still free and that we have neglected the third-order harmonic.

Equation (9.6.1) can also be handled by averaging; we apply the phase-amplitude transformations

$$x(t) = r(t) \cos[t + \phi(t)],$$
$$\dot{x}(t) = -r(t) \sin[t + \phi(t)]$$

to find

$$\dot{r} = -\varepsilon \sin(t + \phi) r^3 \cos^3(t + \phi),$$
$$\dot{\phi} = -\varepsilon \cos(t + \phi) r^2 \cos^3(t + \phi),$$

which averages to

$$\dot{r} = 0, \quad \dot{\phi} = -\varepsilon \tfrac{3}{8} r^2.$$

The result is identical to the one obtained by the harmonic balance method, but we have the additional information that

$$x(t) = R_0 \cos\left(1 - \varepsilon \tfrac{3}{8} R_0^2\right) t \qquad (9.6.3)$$

is an $\mathcal{O}(\varepsilon)$ approximation of a periodic solution of Eq. (9.6.1) on the time scale $1/\varepsilon$ for each R_0, independent of ε.

Example 2

Consider the equation

$$\ddot{x} + x = \varepsilon x^2 \qquad (9.6.4)$$

and substitute again $x(t) = R_0 \cos \Omega t$ into Eq. (9.6.4). We find

$$-R_0 \Omega^2 \cos \Omega t + R_0 \cos \Omega t = \varepsilon R_0^2 \cos^2 \Omega t,$$
$$= \tfrac{1}{2} \varepsilon R_0^2 (1 + \cos 2\Omega t).$$

It is not so clear how to use this as the right-hand side contains a constant and a higher-order harmonic.

We note that averaging produces

$$\dot{r} = 0, \quad \dot{\phi} = 0,$$

so $x(t) = R_0 \cos t$ represents an $\mathcal{O}(\varepsilon)$ approximation on the time scale $1/\varepsilon$ for each R_0.

The one-dimensional nonlinear oscillator with damping and forcing, which was discussed in Section 9.4, can also be analysed with the harmonic balance method and leads to the same results.

The method of harmonic balance is summarised as follows:

- The method requires considerable experience as a prerequisite.
- The method applies to periodic solutions only.
- The method does not include estimates of asymptotic validity.
- The method can be found by use of Galerkin's procedure; this also provides us with numerical error estimates [see Urabe (1965) and Urabe and Reiter (1966)].

9.7 Introduction to Normal Forms

The term normalisation is used whenever an expression or quantity is put in a simpler, standardised form. For instance, an $n \times n$ matrix with constant coefficients can be put in Jordan normal form by a suitable transformation. When all eigenvalues are different, this is a diagonal matrix.

For nonlinear differential equations we can arrive at a normalisation of a vector field in the neighbourhood of a fixed point as follows. Assume that $x = 0$ is the fixed point and write the system of differential equations as

$$\dot{x} = Ax + f(x),$$

where $x \in R^n$ and A is a constant $n \times n$ matrix; $f(x)$ can be expanded in homogeneous vector polynomials, starting with quadratic terms:

$$f(x) = f_2(x) + f_3(x) + \cdots +. \qquad (9.7.1)$$

Here the vector polynomial $f_m(x)$ is homogeneous of degree $m > 2$.

Normalisation of Eq. (9.7.1) means that by successive transformation we remove as many terms of Eq. (9.7.1) as possible. It would be ideal if we could remove all the nonlinear terms, i.e., linearise Eq. (9.7.1) by transformation. In general, however, some nonlinearities will be left.

To normalise Eq. (9.7.1) near $x=0$, we introduce a near-identity transformation of the form

$$x = y + h_2(y) + h_3(y) + \cdots +, \qquad (9.7.2)$$

where $h_m(y)$ are homogeneous vector polynomials of degree m. The unknown polynomials $h_m(y)$ are determined successively by natural requirements as follows. Substituting the near-identity transformation into Eq. (9.7.1) produces

$$\dot{y} + \frac{\partial h_2}{\partial y}\dot{y} + \cdots + = Ay + Ah_2(y) + \cdots + F(y + h_2(y) + \cdots +)$$
$$= Ay + Ah_2(y) + f_2(y) + \cdots, \qquad (9.7.3)$$

where the dots represent cubic and higher-order terms.

Mathematical Methods and Ideas

For $\dot{y}$ we have explicitly

$$\dot{y} = \left[I + \frac{\partial h_2(y)}{\partial y}\right]^{-1} [Ay + Ah_2(y) + f_2(y)] + \cdots +,$$

$$= Ay + Ah_2(y) + f_2(y) - \frac{\partial h_2(y)}{\partial y} Ay + \cdots +, \qquad (9.7.4)$$

where I is the identity operator. It is clear that the quadratic terms vanish if we can find $h_2(y)$ such that

$$\frac{\partial h_2(y)}{\partial y} Ay - Ah_2(y) = f_2(y). \qquad (9.7.5)$$

Equation (9.7.5) is called the homology equation for $h_2(y)$. It is easy to see that for $m > 2$ we have homology equations for $h_m(y)$ of the same form:

$$\frac{\partial h_m(y)}{\partial y} Ay - Ah_m(y) = g_m(y),$$

where $g_m(y)$ are known vector functions. The solvability of the homology equations depends on the invertibility of the linear operator L, which is defined by

$$L(h) = \frac{\partial h}{\partial y} Ay - Ah.$$

Note that the operator L depends on A only; L is invertible if it has no eigenvalues equal to zero.

This observation has important consequences that we explore for the case in which the eigenvalues of the matrix A, $\lambda_1, \ldots, \lambda_n$, are all different. Then we may assume that A is in diagonal form. The homology equation for the term h_m becomes, written out in components $h_{m1}, \ldots, h_{mn}$,

$$\sum_{j=1}^{n} \frac{\partial h_{mi}}{\partial y_j} \lambda_j y_j - \lambda_i h_{mi} = g_{mi},$$

with $m \geq 2$, $i = 1, \ldots, n$.

The terms in h_{mi} are all of the form

$$a y_1^{m_1} y_2^{m_2} \cdots y_n^{m_n} = t_m,$$

Introduction to Normal Forms

where $m_1 + m_2 + \cdots + m_n = m$ and a is a constant. Differentiating, we obtain as the coefficient of term t_m in the homology equation,

$$\sum_{j=1}^{n} m_j \lambda_j - \lambda_i, \quad i = 1, \ldots, n.$$

If this coefficient is zero, we cannot remove t_m. This leads to the following definition.

Definition 1 *The eigenvalues $\lambda_1, \ldots, \lambda_n$ of matrix A are resonant if for some $i \in \{1, 2, \ldots, n\}$ we have*

$$\sum_{j=1}^{n} m_j \lambda_j = \lambda_i,$$

with $m_j \geq 0$ integers and $m_1 + m_2 + \cdots + m_n \geq 2$.

Note that if the eigenvalues of A are nonresonant, we can remove all the nonlinear terms and thereby linearise the system. However, this is less useful than it appears, as in general the sequence of successive transformations to perform this will be divergent.

The usefulness of normalisation lies in removing nonresonant terms to a certain degree to simplify the analysis. We demonstrate this in the following examples.

Example 1
Consider the system

$$\dot{x}_1 = 2x_1 + a_1 x_1^2 + a_2 x_1 x_2 + a_3 x_2^2 + \cdots +,$$
$$\dot{x}_2 = x_2 + b_1 x_1^2 + b_2 x_1 x_2 + b_3 x_2^2 + \cdots +,$$

where the overdots represent polynomial terms of degree three and higher. The matrix A has eigenvalues 2 and 1; the possible resonance relations are

$$2m_1 + m_2 = 2, \quad 2m_1 + m_2 = 1, \quad m_1 + m_2 \geq 2,$$

where m_1 and m_2 are nonnegative integers. The only way to satisfy the resonance relations is for $m_1 = 0$ and $m_2 = 2$. We conclude that, after

167

successive normalisation, only one nonlinear term is left:

$$\dot{y}_1 = 2y_1 + cy_2^2,$$
$$\dot{y}_2 = y_2,$$

where c is a constant that is dependent on the coefficients in the original system. This is a considerable simplification.

The following example is even more important.

Example 2

Consider the perturbed harmonic oscillator:

$$\ddot{x} + \omega^2 x = a_1 x^2 + a_2 x\dot{x} + a_3 \dot{x}^2 + \cdots + . \tag{9.7.6}$$

The eigenvalues of the linear part are $\pm \omega i$, so we have the resonance relations

$$m_1 - m_2 = 1, \quad m_1 - m_2 = -1, \quad m_1 + m_2 \geq 2.$$

These relations imply that after normalisation no even terms are left. On the other hand, an infinite number of odd terms will remain.

One of the interesting consequences is that in 1-degree-of-freedom systems like system (9.7.6) near the equilibrium $(x, \dot{x}) = (0, 0)$ a perturbing odd potential has no effect, locally. This is revealed by the normal form.

To find a first nontrivial effect of the nonlinearities we have to normalise to the cubic terms. An example is given by the van der Pol equation in which the right-hand side of Eq. (9.7.6) is

$$\mu(1 - x^2)\dot{x}.$$

It turns out that, in different coordinates, we find the same results by normalisation as we do by using averaging, described in Sections 9.1–9.3. This is a conclusion that applies quite generally. If both averaging and normalisation can be applied, the resonant terms that are retained are the same for both methods. In general the averaging method is computationally more efficient, but normalisation also covers cases, as in Example 1, in which averaging makes no sense.

For the relation between averaging and normalisation the reader is referred to Sanders and Verhulst (1985) and Verhulst (1996). References for a systematic treatment of normalisation are Arnold (1983) and Golubitsky and Schaeffer (1985).

9.8 Normalisation of Time-Dependent Vector Fields

We have seen periodic time-dependent systems such as the Mathieu equation, oscillators with external forcing, and parametrically excited rotors. Having introduced normalisation in Section 9.7, here we formulate briefly the method and some results for the time-dependent case. Details of proofs and methods to compute the normal-form coefficients can be found in Arnold (1983) and Iooss and Adelmeyer (1992).

Consider the following parameter- and time-dependent equation:

$$\dot{x} = F(x, \mu, t), \tag{9.8.1}$$

with $x \in \mathbf{R}^m$ and the parameters $\mu \in \mathbf{R}^p$. Here $F(x, \mu, t): \mathbf{R}^m \times \mathbf{R}^p \times \mathbf{R} \to \mathbf{R}^m$ is C^∞ in x and μ and T periodic in the t variable. We assume that $x = 0$ is a solution, so $F(0, \mu, t) = 0$, and, moreover, we assume that the linear part of the vector field $D_x F(0, 0, t)$ is time independent for all $t \in \mathbf{R}$. We write $L_0 = D_x F(0, 0, t)$.

Expanding $F(x, \mu, t)$ in a Taylor series with respect to x and μ yields

$$\dot{x} = L_0 x + \sum_{n=2}^{k} F_n(x, \mu, t) + \mathcal{O}[|(x, \mu)|^{k+1}], \tag{9.8.2}$$

where $F_n(x, \mu, t)$ are homogeneous polynomials in x and μ of degree n with T-periodic coefficients.

Theorem 4 *Let $k \in \mathbf{N}$. There exists a (parameter- and time-dependent) transformation $x = \hat{x} + \sum_{n=2}^{k} P_n(\hat{x}, \mu, t)$, where $P_n(\hat{x}, \mu, t)$ are homogeneous polynomials in x and μ of degree n with T-periodic coefficients, such that Eq. (9.8.2) takes the form (with the hat dropped after transformation)*

$$\dot{x} = L_0 x + \sum_{n=2}^{k} \tilde{F}_n(x, \mu, t) + \mathcal{O}[|(x, \mu)|^{k+1}],$$

$$\dot{\mu} = 0. \tag{9.8.3}$$

The truncated vector field,

$$\dot{x} = L_0 x + \sum_{n=2}^{k} \tilde{F}_n(x, \mu, t) = \tilde{F}(x, \mu, t), \tag{9.8.4}$$

which is called the *normal form* of Eq. (9.8.1), has the following properties:

a. $(d/dt)\, e^{L_0^* t} \tilde{F}(e^{-L_0^* t} x, \mu, t) = 0$, for all $(x, \mu) \in \mathbf{R}^{m+p}$, $t \in \mathbf{R}$.
b. If Eq. (9.8.1) is invariant under an involution [i.e., $SF(x, \mu, t) = F(Sx, \mu, t)$, where S is an invertible linear operator such that $S^2 = I$] then the truncated normal-form equation (9.8.4) is also invariant under S. Similarly, if Eq. (9.8.1) is reversible under an involution R [i.e., $RF(x, \mu, t) = -F(Rx, \mu, t)$], then the truncated normal-form equation (9.8.4) is also reversible under R.

Proof. See Iooss and Adelmeyer (1992) and Iooss (1988).

Theorem 4 is applied to situations in which L_0 is semisimple and has only purely imaginary eigenvalues. We take $L_0 = \mathrm{diag}\{i\lambda_1, \ldots, i\lambda_m\}$. In our applications, $m = 2l$ is even and $\lambda_{l+j} = -\lambda_j$ for $j = 1, \ldots, l$. The variable x is then often written as $x = (z_1, \ldots, z_l, \bar{z}_1, \ldots, \bar{z}_l)$. We have the following corollary.

Corollary 2 *Assume that $L_0 = \mathrm{diag}\{i\lambda_1, \ldots, i\lambda_m\}$; then*

a. *A term $x_1^{\gamma_1} \cdots x_m^{\gamma_m} e^{i\frac{2\pi}{T} kt}$ is in the jth component of the Taylor–Fourier series of $\tilde{F}(x, \mu, t)$ if*

$$-\lambda_j + \frac{2\pi}{T} k + \gamma_1 \lambda_1 + \cdots + \gamma_m \lambda_m = 0. \quad (9.8.5)$$

This is known as the resonance condition.
b. *Transforming the normal form through $x = e^{L_0 t} w$ leads to an autonomous equation for w:*

$$\dot{w} = \sum_{n=2}^{k} \tilde{F}_n(w, \mu, 0). \quad (9.8.6)$$

c. *If Eq. (9.8.1) is invariant (respectively reversible) under an involution S, then this also holds for Eq. (9.8.6).*
d. *The autonomous normal-form equation (9.8.6) is invariant under the action of the group $\mathcal{G} = \{g \mid gx = e^{jL_0 T} x,\ j \in \mathbf{Z}\}$,*

generated by $e^{L_0 T}$. *Note that this group is discrete if the ratios of the λ_i are rational and continuous otherwise.*

Proof. See Ruijgrok (1995).

By property b we can make the system autonomous. This is very effective as the autonomous normal-form equation (9.8.6) can be used to prove the existence of periodic solutions and invariant tori of Eq. (9.8.1) near $x = 0$. We then have the following theorem.

Theorem 5 *Let $\varepsilon > 0$, sufficiently small, be given. Scale $w = \varepsilon \hat{w}$.*

a. *If $\hat{w}_0$ is a hyperbolic fixed point of the (scaled) equation (9.8.6), then Eq. (9.8.1) has a hyperbolic periodic solution $x(t) = \varepsilon \hat{w}_0 + \mathcal{O}(\varepsilon^{k+1})$.*
b. *If the scaled equation (9.8.6) has a hyperbolic closed orbit, then Eq. (9.8.1) has a hyperbolic invariant torus.*

Proof. See Sanders and Verhulst (1985), Guckenheimer and Holmes (1983), and Hale (1969) for proofs in the context of averaging. These proofs are easily adapted to the present situation.

These techniques have been applied in Chapters 3, 6, and 7; see also Section 9.10.

9.9 Bifurcations

The new developments in nonlinear dynamics and chaos involve the use of many specialised concepts and terminologies, often with a geometric and abstract flavour. Classically trained scientists have to acquire the basic ideas of these developments to enable them to apply these new powerful insights and methods.

Fortunaly there are many good introductions to nonlinear dynamics available; we mention Guckenheimer and Holmes (1983), Thompson and Stewart (1986), Moon (1987), Wiggins (1990), Verhulst (1996), and the beautiful and readable introduction by Peitgen et al. (1993).

In this section we discuss only the basic ideas of bifurcation theory. Consider a mechanical system described by the n-dimensional system

$$\dot{x} = f(x, t, \mu).$$

The parameter μ, which is m dimensional, represents the natural control parameters of the system, such as friction coefficients, the amplitude of forcing functions, or coupling constants. Such a description of a mechanical system is actually a description of a class of systems, as we will never know the exact values of the parameters μ. Also we are usually interested in what happens for various values of the parameters.

A bifurcation is a qualitative change in the dynamics of the system as the parameter μ crosses a critical value μ_c. In autonomous systems this is reflected by a change of the topology of the phase space. For instance, a saddle with two neighbouring attracting nodes can merge into one attracting node.

Bifurcation theory aims at establishing the critical values μ_c of a system and at describing what happens at the changes. At the value $\mu = \mu_c$ the mechanical system is called structurally unstable; any small change of μ will produce a qualitative change. Sometimes such a change is a small-scale phenomenon, and sometimes it is very dramatic.

9.9.1 Local Bifurcations

The simplest bifurcations to study are the so-called codimension-one bifurcations of equilibria with a one-dimensional parameter. In each case we consider a normal form of the equation near equilibrium, a reason why these are also called local bifurcations.

- The saddle-node bifurcation:

$$\dot{x} = \mu - x^2.$$

If $\mu < 0$ there is no equilibrium solution; at $\mu = 0$ two equilibrium solutions branch off, of which one is stable and one is unstable.
- The transcritical bifurcation:

$$\dot{x} = \mu x - x^2.$$

There are two equilibrium solutions, $x = 0$ and $x = \mu$, with an exchange of stability when μ crosses zero.
- The pitchfork bifurcation:

$$\dot{x} = \mu x - x^3.$$

If $\mu < 0$ there is one equilibrium solution, $x = 0$, that is stable. If $\mu > 0$, there are three equilibrium solutions, $x = 0$ and $x = \pm\sqrt{\mu}$, of which $x = 0$ is unstable and the other two are stable. This pitchfork bifurcation is called supercritical. When $-x^3$ is replaced with $+x^3$, the figure is reflected with respect to the x axis. In this case we call the pitchfork bifurcation subcritical.
- Bifurcation of periodic solutions (*Hopf* bifurcation) can occur when the linearisation of the vector field near an equilibrium has two purely imaginary eigenvalues for a certain value of $\mu+$, whereas the other eigenvalues all have a nonvanishing real part. This situation is well known from the van der Pol equation near the value $\mu = 0$:

$$\ddot{x} + x = \mu(1 - x^2)\dot{x}.$$

It is possible to put the parameter μ in the linear part only by rescaling $u = \sqrt{\mu}x$. For u we find

$$\ddot{u} + u = (\mu - u^2)\dot{u}. \tag{9.9.1}$$

As we know, the equation for x has, if $\mu > 0$, one periodic solution that for small values of μ has the amplitude $2 + O(\mu)$. This means that for u the periodic solution branches off with amplitude $2\sqrt{\mu} + O(\mu^{3/2})$. This behaviour of solutions of the van der Pol equation is typical for Hopf bifurcation.

Averaging is a very efficient technique for handling Hopf bifurcation. A treatment based on polynomial normalisation can be found in Guckenheimer and Holmes (1983).

9.9.2 Global Bifurcations

The bifurcations studied above have the advantage in that we can study them by considering a neighbourhood of an equilibrium or periodic solution. Normal-form calculation or averaging is effective here.

More difficult are the global bifurcations that again take place at certain parameter values, but that are not local in the sense described in Subsection 9.9.1. We make a few introductory remarks in this subsection, but see also Glendinning (1994) and Thomsen (1997) for introductions and Guckenheimer and Holmes (1983) and Wiggins (1990) for more details.

Global bifurcations may cause sudden, large-scale changes in the dynamical system. This is reflected in some of the terminology: catastrophic–explosive or crisis. In autoparametric systems some global bifurcations have been found, for instance, the Šilnikov bifurcations described in Chapter 8. However, general experience with dynamical systems predicts the existence of many more.

As an example we discuss Šilnikov bifurcation briefly. Consider the three-dimensional linear system

$$\dot{x} = \alpha x - \beta y,$$
$$\dot{y} = \beta x + \alpha y,$$
$$\dot{z} = \lambda z, \tag{9.9.2}$$

with $\beta > 0$. The eigenvalues of the equilibrium $(0, 0, 0)$ are $(\alpha \pm i\beta, \lambda)$. If $\alpha, \lambda \neq 0$, this is a saddle focus; if $\lambda > 0$, the axis $x = y = 0$ corresponds with the unstable manifolds and the plane $z = 0$ contains pure spiral motion.

What happens when nonlinear terms are added to system (9.9.2)? Locally near $(0, 0, 0)$ the flow is still rotating and expanding $(\lambda > 0)$ or contracting $(\lambda < 0)$ in the z direction. The invariant plane $z = 0$ is perturbed to a two-dimensional invariant manifold that near $(0, 0, 0)$ is tangent to the plane $z = 0$. Likewise, the axis $x = y = 0$ will be deformed to a one-dimensional stable or unstable manifold that near $(0, 0, 0)$ is tangent to the z axis. Suppose that such a one-dimensional manifold, for a certain parameter value, bends around and ends up again at $(0, 0, 0)$. The solution along the one-dimensional manifold is called homoclinic and the flow near it can be chaotic. We summarise for the Šilnikov bifurcation:

Let X be a sufficiently smooth vector field on $\mathbf{R}^3$ with equilibrium $O = (0, 0, 0)$ such that

1. the eigenvalues of O are $\alpha \pm i\beta$, λ with $|\alpha| < |\lambda|$ and $\beta \neq 0$
2. O has a homoclinic orbit

Then there exists a small perturbation $X + \mu Y$ of the vector field X such that the perturbed field contains a horseshoe map that implies the presence of an infinite number of unstable periodic orbits and chaos.

The difficulty in establishing a Šilnikov bifurcation lies in demonstrating the existence of an homoclinic orbit. In Chapter 7 we use techniques from local bifurcation theory to show the existence of such a homoclinic orbit in a self-excited autoparametric system.

9.10 Bifurcations in a Nonlinear Mathieu Equation

In this section we use the methods developed previously in this chapter to study nonlinear parametric oscillations. As an important example we can think of the nonlinear Mathieu equation.

Consider the following equation:

$$\ddot{x} + k\dot{x} + [\alpha^2 + p(t)]F(x) = 0, \qquad (9.10.1)$$

where $k > 0$ is the damping coefficient, $F(x) = x + bx^2 + cx^3 + \cdots +$, and time is scaled so that

$$p(t) = \sum_{l \in \mathbb{Z}} a_{2l} e^{2ilt}, \quad a_0 = 0, \quad a_{-2l} = \bar{a}_{2l} \qquad (9.10.2)$$

is a π-periodic function with zero average. As we have seen in Section 9.4 [see also Yakubovich and Starzhinskii (1975)], the trivial solution $x = 0$ is unstable when $k = 0$ and $\alpha^2 = n^2$ for all $n \in \mathbb{N}$. Fix a specific $n \in \mathbb{N}$ and assume that α^2 is close to n^2. We study the bifurcations from the solution $x = 0$ in the case of primary resonance, which by definition occurs when the Fourier expansion of $p(t)$ contains nonzero terms $a_{2n}e^{2int}$ and $a_{-2n}e^{-2int}$. The bifurcation parameters in this problem are the detuning $\sigma = \alpha^2 - n^2$, the damping coefficient k, and the Fourier coefficients of $p(t)$, in particular a_{2n}. They are assumed to be small and of equal order of magnitude.

In Broer and Vegter (1992) the conservative case (or Hamiltonian) $k = 0$ was studied; here we consider the dissipative case $k > 0$. The analysis is based on the work of Ruijgrok (1995) and can also be found in Ruijgrok and Verhulst (1996).

9.10.1 Normal-Form Equations

To find the time-periodic normal form of Eq. (9.10.1), we put $x = x_1$ and $\dot{x} = x_2$ and write

$$\dot{x}_1 = x_2,$$
$$\dot{x}_2 = -kx_2 - [n^2 + \sigma + p(t)]F(x_1). \quad (9.10.3)$$

We can write system (9.10.3) in complex form by using $z = nx_1 - ix_2$ and expanding $F(x_1)$:

$$\dot{z} = inz - \tfrac{1}{2}k(z - \bar{z}) + \tfrac{1}{2n}i[\sigma + p(t)](z + \bar{z}) + \cdots +. \quad (9.10.4)$$

The equation for $\bar{z}$ has been omitted.

To Eq. (9.10.4) we apply the time-periodic normal-form procedure as described in the Section 9.8. The right-hand side of Eq. (9.10.4) is expanded in powers of z, $\bar{z}$, and the parameters that are indicated by $\mu = (\sigma, k, a_2, a_4, \ldots,)$. A long calculation yields the time-dependent normal form of Eq. (9.10.4), up to second order:

$$\dot{z} = inz + \left(-\tfrac{1}{2}k + \tfrac{1}{2n}i\sigma\right)z + \tfrac{1}{2n}ia_{2n}e^{i2nt}\bar{z} + \Lambda(z, \bar{z}, \mu, t)$$
$$+ K(z, \bar{z}, \mu, t) + igz|z|^2 + \mathcal{O}[|(z, \bar{z}, \mu)|^4], \quad (9.10.5)$$

where

$$g = \left(\tfrac{3}{4}c - \tfrac{10}{3}b^2\right),$$
$$K(z, \bar{z}, \mu, t) = -\frac{ib}{6n^2}(a_{-n}e^{-int}z^2 + 2a_n e^{int}|z|^2 - 3a_{3n}e^{3int}\bar{z}^2),$$

$$(9.10.6)$$

and $\Lambda(z, \bar{z}, \mu, t)$ contains terms that are linear in z and quadratic in the parameters. It can be assumed that, after a suitable time translation, a_{2n} is real and positive. From this point on, it is assumed that $g \neq 0$. This

Bifurcations in a Nonlinear Mathieu Equation

condition is satisfied by almost all choices of $F(x)$ and can therefore be called generic.

The normal-form equation (9.10.5) can be made autonomous through the transformation $z = w e^{int}$. After time is scaled with a factor $1/2n$ and $\kappa = nk$ is introduced, the equation for w becomes

$$\dot{w} = (-\kappa + i\sigma)w + ia_{2n}\bar{w} + K(w, \bar{w}, \mu) + igw|w|^2$$
$$+ \hat{\Lambda}(w, \bar{w}, \mu) + \mathcal{O}[|(w, \bar{w}, \mu)|^4], \qquad (9.10.7)$$

where now

$$K(w, \bar{w}, \mu) = -\frac{ib}{3n}(a_{-n}w^2 + 2a_n|w|^2 - 3a_{3n}\bar{w}^2), \quad (9.10.8)$$

$$\hat{\Lambda}(w, \bar{w}, \mu) = 2nL(w, \bar{w}, \mu, 0). \qquad (9.10.9)$$

We now scale

$$\sigma = \varepsilon\hat{\sigma}, \quad \kappa = \varepsilon\hat{\kappa}, \quad a_{2j} = \varepsilon\hat{a}_{2j}, \quad j = 1, 2, \ldots,. \quad (9.10.10)$$

Following Broer and Vegter (1992) [in which it is shown that all nontrivial fixed points of Eq. (9.10.7) are at $\mathcal{O}(\varepsilon^{1/2})$ of the origin], we also scale $w = \varepsilon^{1/2}\hat{w}$. Equation (9.10.7) becomes (with the hats dropped and time scaling $\tau = \varepsilon t$)

$$w' = (-\kappa + i\sigma)w + ia_{2n}\bar{w} + \varepsilon^{1/2}K(w, \bar{w}, \mu) + igw|w|^2 + \mathcal{O}(\varepsilon),$$
$$(9.10.11)$$

where $K(w, \bar{w}, \mu)$ is as given in Eq. (9.10.8). Note that the $\mathcal{O}(\varepsilon)$ estimate is valid, as it is easy to see that even terms in $(z, \bar{z})$ have coefficients of $\mathcal{O}(\varepsilon)$, so that, in particular, terms of degree 4 will lead to $\mathcal{O}(\varepsilon)$ terms in rescaled equation (9.10.11).

Equation (9.10.11) is invariant under $(w, \bar{w}) \to -(w, \bar{w})$ [up to $\mathcal{O}(\varepsilon)$ terms] if and only if $K(w, \bar{w}, \mu) = 0$. For sufficiently small ε, Eq. (9.10.11) can be treated as a perturbation of a symmetric system. This motivates us to study first the system with symmetric terms.

9.10.2 Dynamics and Bifurcations of the Symmetric System

In this section we assume that the autonomous normal-form equation (9.10.7) is invariant under $(w, \bar{w}) \to -(w, \bar{w})$. This symmetry implies,

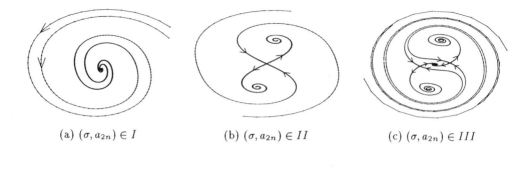

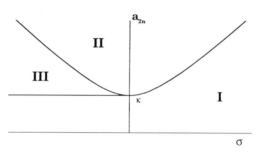

Figure 9.3: Bifurcation diagram in the (σ, a_{2n}) plane and phase-portraits of Eq. (9.10.12).

among other things, that all fixed points come in pairs and that bifurcations of the origin will be symmetric (such as pitchfork bifurcations). We observe that the normal-form equation is symmetric when either $F(x)$ in Eq. (9.10.1) is odd in x or when n is odd. This is reflected by Eq. (9.10.11), which is invariant under $(x, y) \rightarrow -(x, y)$ only if the quadratic terms vanish. From Eq. (9.10.8) it is easy to see that this indeed is the case when $F(x)$ is odd, because then $b = 0$. Similarly, when n is odd, $p(t) = \sum_{l \in \mathbb{Z}} a_{2l} e^{2ilt}$ does not contain terms a_{-n}, a_n, or a_{3n}, and all the coefficients in Eq. (9.10.8) equal zero.

The symmetric equation, truncated at $\mathcal{O}(\varepsilon)$, is given by

$$\dot{w} = (-\kappa + i\sigma)w + ia_{2n}\bar{w} + igw|w|^2. \qquad (9.10.12)$$

It is not difficult to show that, for sufficiently large R, the disk $|w| < R$ is invariant under the flow of Eq. (9.10.12) and that the only attractors in this region are fixed points. The dynamics of Eq. (9.10.12) are summarised in Figure 9.3.

Outside the hyperbola $\kappa^2 + \sigma^2 = a_{2n}^2$ (that is, outside region II), the trivial solution is stable. On the hyperbola a pitchfork bifurcation occurs, which is supercritical if $\sigma > 0$ and subcritical if $\sigma < 0$. On the half line $a_{2n} = \kappa$, $\sigma < 0$ there occurs a double saddle-node bifurcation, i.e., two simultaneous saddle nodes.

9.10.3 Bifurcations in the General Case

As was remarked at the end of Subsection 9.10.1, the general equation (9.10.11) can be seen as a nonsymmetric $\mathcal{O}(\varepsilon^{1/2})$ perturbation of the symmetric case. For most values of the parameters σ and a_{2n}, the phase portraits of the symmetric equation are structurally stable, so for sufficiently small ε, the perturbation will have no qualitative effect. There will still be zero, two, or four (nontrivial) fixed points, and they will remain hyperbolic for values of (σ, a_{2n}) outside a neighbourhood of the boundaries in the bifurcation diagram (Figure 9.3). These fixed points will, however, no longer come in symmetric pairs. For the half line $a_{2n} = \kappa$, $\sigma < 0$ in Figure 9.3, we can make the following remark. In the symmetric case, two saddle-node bifurcations occur simultaneously. Because saddle-node bifurcations are generic, they will persist in the perturbed case. However, because of the symmetry breaking, they will, in general, no longer occur simultaneously. We therefore expect that the half line will break up into two curves of saddle-node bifurcations. These considerations hold outside a neighbourhood of the point $(\sigma, a_{2n}) = (0, \kappa)$. Near this point we will find more complicated behaviour.

It follows that we have to consider values of (σ, a_{2n}) only near the hyperbola $\kappa^2 + \sigma^2 = a_{2n}^2$. For these values of the parameters (that is, at points in parameter space where the trivial solution loses stability), we then apply the centre-manifold theory. Let $\lambda = -\kappa^2 - \sigma^2 + a_{2n}^2$ be the bifurcation parameter. In Ruijgrok (1995) it is shown that the flow in the centre manifold is given by

$$\dot{u} = \mu_1 u + \varepsilon^{1/2} \mu_2 u^2 + \mu_3 u^3 + \varepsilon^{1/2} \mu_4 u^4 - u^5,$$
$$\dot{\lambda} = 0, \qquad (9.10.13)$$

where μ_1 is proportional to λ and μ_3 is proportional to σ. The coefficients μ_2 and μ_4 are functions of λ, μ_3, and the Fourier coefficients a_n and

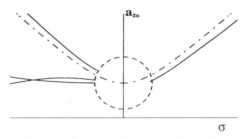

Figure 9.4: Partial bifurcation diagram in the (σ, a_{2n}) plane in the nonsymmetric case.

a_{3n}. The most degenerate member of the family of equations (9.10.13) is $\dot{u} = -u^5$, and it is not difficult to see that Eqs. (9.10.13) define a four-parameter unfolding of this degeneracy, such that $u = 0$ is always a solution [see Ruijgrok (1995) for details].

If σ is not small, the term u^3 in Eqs. (9.10.13) dominates. We then have a symmetry-breaking perturbation of the pitchfork bifurcation [see Golubitsky and Schaeffer (1985)] that leads to a saddle-node bifurcation followed (or preceded) by a transcritical bifurcation.

We can now sketch the following partial bifurcation diagram in (σ, a_{2n}) space, excluding a neighbourhood of the point $(\sigma, a_{2n}) = (0, \kappa)$ (see Figure 9.4).

It remains to analyse the case in which σ is small. Note that then μ_3 is also small. Rescale λ and σ through $\lambda = \varepsilon^{1/2} \hat{\lambda}$, $\sigma = \varepsilon^{1/2} \hat{\sigma}$. Truncating at $\mathcal{O}(\varepsilon^{1/2})$ and, as always, dropping the hats yield the following equation for u:

$$\dot{u} = \mu_1 u + \varepsilon^{1/2} \mu_2 u^2 + \mu_3 u^3 + \varepsilon^{1/2} \mu_4 u^4 - u^5. \quad (9.10.14)$$

The translation $U = u + \frac{1}{5}\mu_4$ takes Eq. (9.10.14) to

$$\dot{U} = v_1 + v_2 U + v_3 U^2 + v_4 U^3 - U^5, \quad (9.10.15)$$

where v_i, $i = 1, \ldots, 4$ are functions of μ_j. The bifurcation set of this equation (the "butterfly") is thoroughly examined in Poston and Stewart (1978) and Broecker and Lander (1975). The bifurcation set in $(\mu_1, \mu_2, \mu_3, \mu_4)$ space is the image of the butterfly under the smooth transformation $(v_1, v_2, v_3, v_4) \to (\mu_1, \mu_2, \mu_3, \mu_4)$.

A difficult problem is how to depict the bifurcation set, as it lives in a four-dimensional space. Following an idea in Broecker and Lander

(1975), we give a series of bifurcation pictures in the (μ_1, μ_3) plane as the values of (μ_2, μ_4) are varied (see Figures 9.5 and 9.6). We choose (μ_1, μ_3), as we are interested in completing the bifurcation diagram of Figure 9.4. The line $\mu_1 = 0$ corresponds to the hyperbola $\kappa^2 + \sigma^2 = a_{2n}^2$, and the line $\mu_3 = 0$ corresponds to the line $\sigma = 0$ (in Figure 9.5, we have actually reversed the direction of μ_3, so that now positive μ_3 corresponds to positive σ). It is therefore easy to transform a bifurcation diagram in the (μ_1, μ_3) plane to one in the (σ, a_{2n}) plane. As an example, consider Figure 9.5, which shows the bifurcation diagram in the (μ_1, μ_3) plane for $\mu_2 = 0$ and $\mu_4 = 0$. This diagram, representing the symmetric case, is equivalent to Figure 9.3.

In this way, it is not difficult to complete the bifurcation diagram of Figure 9.4, and the result is shown in Figure 9.7. Note that there are three topologically distinct possibilities.

9.10.4 Discussion

Consider again the original equation (9.10.4), written in complex form,

$$\dot{z} = F(z, \bar{z}, \mu, t), \quad z \in \mathbf{C}, \tag{9.10.16}$$

and its normal form,

$$\dot{\tilde{z}} = \tilde{F}(\tilde{z}, \bar{\tilde{z}}, \mu, t), \quad z \in \mathbf{C}, \tag{9.10.17}$$

where $F(., t)$ and $\tilde{F}(., t)$ are π periodic. The variables z and $\tilde{z}$ are related through $z = \tilde{z} + h(\tilde{z}, \bar{\tilde{z}}, \mu)$ with $h(\tilde{z}, \bar{\tilde{z}}, \mu)$, a C^∞ function whose Taylor expansion starts with quadratic terms.

We can characterise the dynamics of systems (9.10.16) and (9.10.17) completely by studying suitable associated maps. An important tool in this is the Poincaré map, which is relatively easy to construct, also numerically. For an extensive discussion of such maps and how to use them, see Guckenheimer and Holmes (1983) and Verhulst (1996). An additional bonus is that many theorems in dynamical systems theory have been formulated for maps; in our discussion we can immediately apply these theorems. We study the dynamics of Eq. (9.10.16) through the Poincaré map $P_\mu : \mathbf{C} \to \mathbf{C}$, which is defined by

$$P_\mu(z) = X_\pi^\mu(z, 0), \tag{9.10.18}$$

where $X_t^\mu(v, t_0)$ is the solution of Eq. (9.10.16) with initial conditions

Mathematical Methods and Ideas

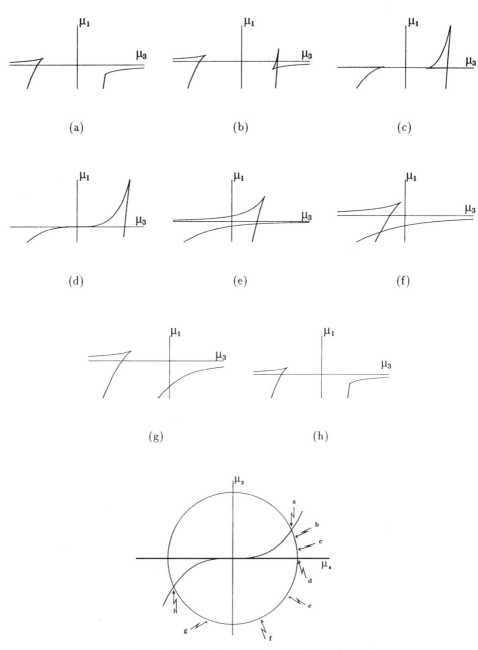

Figure 9.5: Bifurcation diagram in the (μ_1, μ_3) plane.

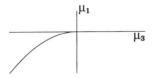

Figure 9.6: Bifurcation diagram in the (μ_1, μ_3) plane in the symmetric case $(\mu_2 = \mu_4 = 0)$.

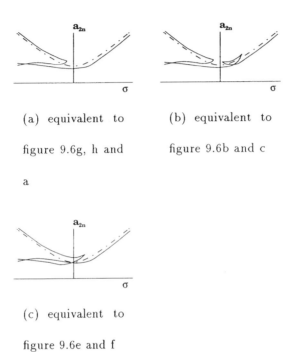

(a) equivalent to figure 9.6g, h and a

(b) equivalent to figure 9.6b and c

(c) equivalent to figure 9.6e and f

Figure 9.7: Bifurcation diagrams in the (a_{2n}, σ) plane.

$z(t_0) = v$ and $\bar{z}(t_0) = \bar{v}$. The Poincaré map of Eq. (9.10.17) is denoted by $\tilde{P}_\mu$.

A well-known theorem on normal forms [see for instance Arnold (1983)] implies that if $\tilde{z}_0$ is a hyperbolic fixed point of $\tilde{P}_\mu(\tilde{z})$, then there exists a nearby hyperbolic fixed point z_0 (of the same stability type) of $P_\mu(z)$. In this case, $z_0 = \tilde{z}_0 + \mathcal{O}(\varepsilon^{1/2})$.

Recall that the normal-form equation (9.10.17) is made autonomous through the transformation

$$z = e^{int} w, \qquad (9.10.19)$$

which leads to

$$\dot{w} = \tilde{F}(w, \bar{w}, \mu, 0). \qquad (9.10.20)$$

Now suppose that w_0 is a hyperbolic fixed point of Eq. (9.10.20). The image of w_0 under the Poincaré map $\tilde{P}_\mu$ is then

$$\tilde{P}_\mu(w_0) = e^{in\pi} w_0, \qquad (9.10.21)$$

where we have used Eqs. (9.10.18) and (9.10.19).

From Eq. (9.10.21) it follows that, if n is even, w_0 is a hyperbolic fixed point of $\tilde{P}_\mu$, corresponding to a hyperbolic fixed point of P_μ, corresponding to a hyperbolic π-periodic solution of Eq. (9.10.16). However, when n is odd, we see that $\tilde{P}_\mu(w_0) = -w_0$ and that $(w_0, -w_0)$ is a hyperbolic orbit of period 2 of $\tilde{P}_\mu$, which corresponds to a hyperbolic 2π-periodic solution of Eq. (9.10.16).

The bifurcations occurring in the autonomous normal-form equation (9.10.20) have the following implications for the original equation (9.10.16). If n is odd, the pitchfork bifurcation of the origin in Eq. (9.10.20) corresponds to a period-doubling (or flip) bifurcation of the solution $z = 0$ of the Poincaré map of Eq. (9.10.16). The normal form of the flip bifurcation is given by the one-dimensional map $F(u, \lambda) = -u + \lambda u \pm u^3$ [see Wiggins (1990)] in which the sign of the cubic term determines whether this bifurcation is supercritical or subcritical. In general, a pair of fixed points $(w_0, -w_0)$ of Eq. (9.10.20) corresponds to one 2π-periodic solution of Eq. (9.10.16). Therefore, for parameter values in regions II and III in Figure 9.3, there are 1 and 2, respectively, 2π-periodic orbits. The symmetric saddle-node bifurcation corresponds with one saddle-node bifurcation of 2π-periodic orbits.

If n is even but the function $F(x)$ in Eq. (9.10.16) is odd in x, the normal-form equation (9.10.20) is the same as in the preceding case. However, in this case the pitchfork bifurcation of the origin corresponds to a pitchfork bifurcation of π-periodic orbits in Eq. (9.10.16). The normal form of the pitchfork bifurcation is given by the one-dimensional map $F(u, \lambda) = u + \lambda u \pm u^3$. For parameter values in regions II and III in Figure 9.3, there are 2 and 4, respectively, π-periodic orbits. The symmetric saddle-node bifurcation corresponds to two saddle-node bifurcations of π-periodic orbits occurring simultaneously. If n is even and $F(x)$ is not odd in x, there are only transcritical and saddle-node bifurcations occurring in Eq. (9.10.20), corresponding to transcritical and saddle-node bifurcations of π-periodic orbits. The normal forms are $F(u, \lambda) = u + \lambda u \pm u^2$ and $F(u, \lambda) = u + \lambda \pm u^2$, respectively.

In Broer and Vegter (1992), the Hamiltonian case ($k=0$) of Eq. (9.10.1) is studied. Comparing it with this study, we find similarities and differences.

The similarity is in the autonomous normal-form equation (9.10.7), which for $k = 0$ is the same as that calculated in Broer and Vegter (1992). Also, to the special (symmetric) cases of this section, there correspond similar symmetric Hamiltonian cases, with similar codimension-one bifurcations (transcritical, saddle node, pitchfork, or flip, depending on the specific symmetry). In the Hamiltonian case, however, there is an additional possibility of symmetry, namely, when Eq. (9.10.1) is time reversible [i.e., when $p(t) = p(-t)$ for all $t \in \mathbf{R}$]. Reversibility does not occur in the present dissipative case.

The most important difference is that in the dissipative case, the bifurcation analysis can be reduced to the study of a one-dimensional system. Also, there are few difficulties in translating the bifurcation results for the autonomous normal form back to the original equation. The Hamiltonian case is rather more complicated. In Broer and Vegter (1992), the authors use the singularity theory for families of planar Hamiltonian functions in their analysis. They reduce the normal form to a two-parameter family of Hamiltonians (these parameters are roughly equivalent with σ and a_{2n}). In general, the analysis in the Hamiltonian case is more subtle than in the dissipative case.

9.11 The Mathieu Equation with Nonlinear Damping

In this section we consider a different type of nonlinear Mathieu equation, in which now the damping provides the nonlinear term. The results are well known, but as there is no adequate survey available, we summarise the results in three significant cases. Each of them contains saddle-node bifurcations that lead to stable periodic solutions.

1. Quadratic damping $|x|\dot{x}$:
    The equation becomes
    $$\ddot{x} + (1 + 4\varepsilon\eta^2 \cos 2\eta t)x + \varepsilon\kappa\dot{x} + \varepsilon\delta f(x, \dot{x})\dot{x} = 0, \quad (9.11.1)$$

with $\kappa > 0$, $\delta \geq 0$; first we take $f(x, \dot{x}) = |\dot{x}|$, so we have progressive nonlinear damping. Putting again $\tau = \eta t$, we have

$$x'' = (1/\eta^2 + 4\varepsilon \cos 2\tau)x + \varepsilon \frac{\kappa}{\eta} x' + \varepsilon \frac{\delta}{\eta} |x'|x' = 0, \quad (9.11.2)$$

with primary resonance if η is near to 1:

$$\eta = 1 + \varepsilon \sigma. \quad (9.11.3)$$

To perform averaging [see Verhulst (1996)], we use an amplitude-phase representation $x = r \cos(\tau + \psi)$ and $x' = -r \sin(\tau + \psi)$, and to study the trivial equilibrium solution, we use Cartesian coordinates $u = r \cos \psi$ and $v = r \sin \psi$; note that we are not permitted to use polar coordinates near $(0, 0)$.

Introducing the Lagrange standard form, averaging, and omitting the terms of the order of ε^2, we find

$$r' = \varepsilon \left(r \sin 2\psi - \tfrac{1}{2}\kappa r - \tfrac{1}{2}\delta r^2 \right),$$
$$\psi' = \varepsilon(-\sigma + \cos 2\psi). \quad (9.11.4)$$

In Cartesian coordinates, system (9.11.4) becomes

$$u' = \varepsilon \left[-\tfrac{1}{2}\kappa u + (1+\sigma)v - \tfrac{1}{2}\delta(u^2+v^2)^{1/2} u \right],$$
$$v' = \varepsilon \left[-\tfrac{1}{2}\kappa v + (1-\sigma)u - \tfrac{1}{2}\delta(u^2+v^2)^{1/2} v \right]. \quad (9.11.5)$$

Apart from the trivial solution $(u, v) = (0, 0)$, two nontrivial equilibrium solutions exist if

$$\sigma^2 < 1 - \tfrac{1}{4}\kappa^2, \quad \delta > 0. \quad (9.11.6)$$

Solution $(0, 0)$ is unstable (two real eigenvalues, one positive) if inequalities (9.11.6) are satisfied; at $\sigma^2 = 1 - \tfrac{1}{4}\kappa^2$ the nontrivial solutions vanish and $(0, 0)$ becomes asymptotically stable (two negative eigenvalues) for increasing σ.

At the two points $\sigma \doteq \pm(1-\tfrac{1}{4}\kappa^2)^{1/2}$ we have a saddle-node bifurcation that produces a stable nontrivial equilibrium solution. Returning to the original coordinates x and x', we find that the equilibrium solutions presented here correspond with 2π-periodic solutions in τ of the original system.

Near the primary resonance $\eta = 1$ the basic motion becomes unstable but, because of progressive nonlinear damping, the motion remains bounded and tends towards a stable periodic solution. The $O(\varepsilon)$ approximations valid for all time are

$$x_p(t) = \frac{2\sqrt{1-\sigma^2}-\kappa}{\delta} \cos(2\eta t + \psi_p),$$

$$\cos 2\psi_p = \sigma, \quad \sin 2\psi_p > 0, \quad (9.11.7)$$

with two solutions for ψ_p.

2. Cubic damping $\dot{x}^3$:

In Eq. (9.11.1) we take $f(x, \dot{x}) = \dot{x}^2$. When averaging is performed to first order near the primary resonance of Eq. (9.11.3), only the terms with coefficients δ are changing. In Eqs. (9.11.4) the last term in the equation for r becomes $-\frac{3}{8}\delta r^3$. In Eqs. (9.11.5) the last terms become $-\frac{3}{8}\delta(u^2+v^2)u$ and $-\frac{3}{8}\delta(u^2+v^2)v$.

Again we have that, apart from the trivial solution $(u, v) = (0, 0)$, two nontrivial equilibrium solutions exist if

$$\sigma^2 < 1 - \frac{1}{4}\kappa^2, \quad \delta > 0.$$

The stability behaviour of the equilibrium solutions is exactly as in the case of quadratic damping. If the nontrivial equilibrium solutions exist, $O(\varepsilon)$ approximations of the corresponding periodic solution, valid for all time, are

$$x_p(t) = 2\left(\frac{2\sqrt{1-\sigma^2}-\kappa}{3\delta}\right)^{1/2} \cos(2\eta t + \psi_p),$$

$$\cos 2\psi_p = \sigma, \quad \sin 2\psi_p > 0, \quad (9.11.8)$$

with two solutions for ψ_p.

3. Cubic damping $x^2\dot{x}$:

In Eq. (9.11.1) we now have $f(x, \dot{x}) = x^2$. When averaging is performed near the primary resonance of Eq. (9.11.3) again, only the terms with coefficients δ are changing. In Eqs. (9.11.4) the last term in the equation for r becomes $-\frac{1}{8}\delta r^3$. The bifurcation behaviour, inequality (9.11.6), and the stability behaviour are the same as before. If the nontrivial equilibrium solutions exist, $O(\varepsilon)$ approximations of the corresponding periodic solution, valid for all time, are

$$x_p(t) = 2\left(\frac{2\sqrt{1-\sigma^2} - \kappa}{\delta}\right)^{1/2} \cos(2\eta t + \psi_p),$$

$$\cos 2\psi_p = \sigma, \quad \sin 2\psi_p > 0, \qquad (9.11.9)$$

with two solutions for ψ_p.

Bibliography

Arnold, V.I., *Geometrical Methods in the Theory of Ordinary Differential Equations*, Springer-Verlag, New York, 1983.

Bajaj, A.K., Chang, S.I., and Johnson, J.M., Amplitude modulated dynamics of a resonantly excited autoparametric two degree of freedom system, *Nonlinear Dyn.* **5**, 433–457, 1994.

Banerjee, B., Bajaj, A.K., and Davies, P., Second-order averaging study of an autoparametric system, in *Proceedings of the Fourteenth Biennial ASME Conference on Mechanical Vibration and Noise,* American Society of Mechanical Engineers, New York, 1993.

Banerjee, B., Bajaj, A.K., and Davies, P., Resonant dynamics of an autoparametric system: A study using higher order averaging, *Int. J. Non-Linear Mech.* **31**, 21–39, 1996.

Banichuk, N.V., Bratus, A.S., and Myshkis, A.D., Stabilizing and destabilizing effects in nonconservative systems, *PMM USSR* **53**(2), 158–164, 1989.

Bogoliubov, N.N. and Mitroplosky, Yu. A., *Asymptotic Methods in the Theory of Nonlinear Oscillations*, Gordon & Breach, New York, 1961.

Bolotin, V.V., *Non-Conservative Problems of the Theory of Elastic Stability*, Pergamon, Oxford, 1963.

Bratus, A.S., On static stability of elastic nonconservative systems with small damping, Preprint, Moscow Railway Engineering Institute, Moscow, 1990.

Broecker, Th. and Lander, L., *Differentiable Germs and Catastrophes*, Vol. 17 of London Mathematical Society Lecture Notes Series, Vol. 17, Cambridge U. Press, New York, 1975.

Broer, H.W. and Vegter, G., Subordinate Silnikov bifurcations near some singularities having low codimension, *Ergodic Theory Dyn. Syst.* **4**, 509–525, 1984.

Bibliography

Broer, H.W. and Vegter, G., Bifurcational aspects of parametric resonance, in *Dynamics Reported: New Series Vol. 1: Expositions in Dynamical Systems*, C.K.R.T. Jones, U. Kirchgraber, and H.O. Walther, eds., Springer, New York, 1992.

Cartmell, M., *Introduction to Linear, Parametric and Nonlinear Vibrations*, Chapman & Hall, London, 1990.

Dankowicz, H., *Chaotic Dynamics in Hamiltonian Systems with Applications to Celestial Mechanics*, Vol. 25 of Nonlinear Science Series A, World Scientific, Singapore, 1997.

Doedel, E.J., AUTO: A program for the automatic bifurcation analysis of autonomous systems, *Congr. Numer.* **30**, 265–284, 1981.

Froude, W., On the rolling of ships, *Trans. Inst. Nav. Archit.* **2**, 180–229, 1861.

Glendinning, P., *Stability, Instability and Chaos*, Cambridge U. Press, New York, 1994.

Golubitsky, M. and Schaeffer, D., *Singularities and Groups in Bifurcation Theory*, Springer-Verlag, New York, 1985.

Grim, O., Rollschwingungen, Stabilität und Sicherheit im Seegang, *Schiffstechnik* **1**, 10–21, 1952.

Guckenheimer, J., On a codimension two bifurcation, in *Dynamical Systems and Turbulence*, Warwick, Springer LNM 898, Springer-Verlag, New York, 1980.

Guckenheimer, J. and Holmes, P., *Nonlinear Oscillations, Dynamical Systems and Bifurcations of Vectorfields*, Springer-Verlag, New York, 1983.

Hale, J., *Ordinary Differential Equations*, Wiley, New York, 1969.

Hatwal, H., Mallik, A.K., and Ghosh, A., Forced nonlinear oscillations of an autoparametric system, *Am. Soc. Mech. Eng. J. Appl. Mech.* **50**, 657–662, 1983.

Haxton, R.S. and Barr, A.D.S., The autoparametric vibration absorber, *Am. Soc. Mech. Eng. J. Eng. Ind.* **94**, 119–125, 1972.

Hoveijn, I. and Ruijgrok, M., The stability of parametrically forced coupled oscillators in sum resonance, *Z. Angew. Math. Phys.* **46**, 383–392, 1995.

Iooss, G., Global characterisation of the normal form for a vectorfield near a closed orbit, *J. Diff. Eqns.* **76**, 47–76, 1988.

Iooss, G. and Adelmeyer, M., *Topics in Bifurcation Theory*, World Scientific, Singapore, 1992.

Kerwin, J.E., Notes on rolling in longitudinal waves, *Int. Shipbuild. Prog.* **16**, 597–614, 1955.

Klein, F. and Sommerfeld, A., *Über die Theory des Kreisels*, B.G. Teubner Verlags Gesellschaft, Stuttgart, Germany 1897 ed., 1965.

Korvin-Kroukovsky, B.V., *Theory of Seakeeping*, Society of Naval Architects and Marine Engineers, New York, 1961.

Mook, D.T., Marshall, L.R., and Nayfeh, A.H., Subharmonic and superharmonic resonances in the pitch and roll modes of ship motions, *J. Hydronaut.* **8**, 32–40, 1974.

Moon, F.C., *Chaotic Vibrations: An Introduction for Applied Scientists and Engineers*, Wiley, New York, 1987.

Nabergoj, R. and Tondl, A., Simulation of parametric ship rolling: Effects of hull bending and torsional elasticity, *Nonlinear Dyn.* **6**, 265–284, 1994.

Nabergoj, R. and Tondl, A., Autoparametric resonance in a self-excited single-mass system, in *Proceedings of EUROMECH—2nd European Nonlinear Oscillation Conference*, Institute of Thermomechanics, Academy of Sciences of the Czech Republic, Prague, 1996, Vol. 2, pp. 151–156.

Nabergoj, R., Tondl, A., and Virag, Z., Autoparametric resonance in an externally excited system, *Chaos, Solitons Fractals* **4**, 263–273, 1994.

Nayfeh, A.H. and Mook, D.T., *Nonlinear Oscillations*, Wiley Interscience, New York, 1979.

Nayfeh, A.H., Mook, D.T., and Marshall, L.R., Nonlinear coupling of pitch and roll modes in ship motions, *J. Hydronaut.* **4**, 145–152, 1973.

Newhouse, S.E., Ruelle, D., and Takens, F., Occurrence of strange axiom A attractors near quasi-periodic flow on T^m, $m \leq 3$, *Commun. Math. Phys.* **64**, 1978.

Paulling, J.R., The transverse stability of a ship in a longitudinal seaway, *J. Ship Res.* **5**, 37–49, 1961.

Paulling, J.R. and Rosenberg, R.M., On unstable ship motions resulting from nonlinear coupling, *J. Ship Res.* **3**, 36–46, 1959.

Peitgen, H.O., Jürgens, H., and Saupe, D., *Chaos and Fractals*, Springer-Verlag, New York, 1993.

Poston, T. and Stewart, I., *Catastrophe Theory*, Pitman, London, 1978.

Roseau, M., *Vibrations Nonlinéaires et Théorie de la Stabilité*, Springer-Verlag, New York, 1966.

Ruijgrok, M., Studies in parametric and autoparametric resonance, Thesis, Utrecht University, Utrecht, The Netherlands, 1995.

Ruijgrok, M. and Verhulst, F., Parametric and autoparametric resonance, *Prog. Nonlinear Diff. Eqns. Appl.* **19**, 279–298, 1996.

Ruijgrok, M., Tondl, A., and Verhulst, F., Resonance in a rigid rotor with elastic support, *Z. Angew. Math. Mech.* **73**, 255–263, 1993.

Sanders, J.A. and Verhulst, F., *Averaging Methods in Nonlinear Dynamical Systems*, Vol. 59 of Applied Mathematical Sciences, Springer-Verlag, New York, 1985.

Schmidt, G., *Parametererregte Schwingungen*, VEB Deutscher Verlag der Wissenschaften, Berlin, 1975.

Schmidt, G. and Tondl, A., *Non-Linear Vibrations*, Akademie-Verlag, Berlin, 1986.

Šilnikov, L.P., A case of the existence of a denumerable set of periodic motions, *Sov. Math. Dokl.* **6**, 163–166, 1965.

Svoboda, R., Tondl, A., and Verhulst, F., Autoparametric resonance by coupling of linear and nonlinear systems, *Int. J. Non-linear Mech.* **29**, 225–232, 1994.

Szemplinska-Stupnicka, W., *The Behaviour of Nonlinear Vibrating Systems*, Kluwer Academic, Dordrecht/Boston/London, 1990, Vol. II.

Takens, F., *Singularities of Vectorfields*, Publ. Math. Institut des Hantes Études Scientifiques Bures-sur-Yuette (France), 1974, Vol. 43, pp. 47–100.

Thompson, J.M.T. and Schiehlen, W., Nonlinear dynamics of engineering systems, *Philos. Trans. R. Soc. London, Ser. A* **338**, 1992.

Thompson, J.M.T. and Stewart, I., *Nonlinear Dynamics and Chaos*, Wiley, New York, 1986.

Thomsen, J.J., Chaotic vibrations of non-shallow arches, *J. Sound Vib.* **153**, 239–258, 1992.

Thomsen, J.J., *Vibrations and Stability*, McGraw-Hill, London, 1997.

Tondl, A., A new model of a self-controlled system, *Acta Tech. Cesk. Akad. Ved* **32**, 243–256, 1987.

Tondl, A., A model of a self-controlled system with hysteresis, *Acta Tech. Cesk. Akad. Ved* **33**, 269–281, 1988a.

Tondl, A., Dynamic study of a rigid rotor elastically supported in axial and lateral directions, Lecture Notes, Politecnico di Torino, Torino, 1988b.

Tondl, A., On the stability of a rotor system, *Acta Tech. Cesk. Akad. Ved* **36**, 331–338, 1991a.

Tondl, A., *Quenching of Self-Excited Vibrations*, Elsevier, Amsterdam, 1991b.

Tondl, A., A contribution to the analysis of an autoparametric system, in *Festschrift zum 80 Geburtstag von Prof. K. Magnus*, Technische Universität München, Munich, 1992, pp. 291–304.

Tondl, A., Parametric resonance vibration in a rotor system, *Acta Tech. Cesk. Akad. Ved* **37**, 185–194, 1992a.

Tondl, A., A contribution to the analysis of autoparametric systems, *Acta Tech. Cesk. Akad. Ved* **37**, 735–758, 1992b.

Tondl, A., Elastically mounted body in cross flow with an attached pendulum, in *Proceedings of the Fourteenth Biennial ASME Conference on Mechanical Vibration and Noise, Dynamics and Vibration of Time-Varying Systems and Structures*, American Society of Mechanical Engineers, New York, 1993, pp. 93–96.

Tondl, A., Fiala, V., and Šklíba, J., Domains of attraction for nonlinear systems, Monographs and Memoranda No. 8, National Research Institute for Machine Design, Běchovice, 1970.

Tondl, A., Kotek, V., and Kratochvil, C., Analysis of an autoparametric system, in *Proceedings of EUROMECH—2nd European Nonlinear Oscillation Conference*, Institute of Thermomechanics, Academy of Sciences of the Czech Republic, Prague, 1996, Vol. 1, pp. 467–470.

Tondl, A. and Nabergoj, R., Model simulation of parametrically excited ship rolling, *Nonlinear Dyn.* **1**, 131–141, 1990.

Tondl, A. and Nabergoj, R., Simulation of parametric ship hull and twist oscillations, *Nonlinear Dyn.* **3**, 41–56, 1992.

Tondl, A. and Nabergoj, R., Autoparametric systems, Rep. 16, University of Trieste, Department of Naval Architecture, Ocean and Enviromental Engineering, Trieste, Italy, 1993.

Tondl, A. and Nabergoj, R., Non-periodic and chaotic vibrations in a flow induced system, *Chaos, Solitons Fractals* **4**, 2193–2202, 1994.

Tondl, A. and Nabergoj, R., Autoparametric systems with pendula, in *Wave Motion, Intelligent Structures and Nonlinear Mechanics*, Vol. 1 of Stability, Vibration and Control of Structures, Series, A. Guran and D.J. Inman, eds., World Scientific, Singapore, 1995, pp. 226–238.

Tondl, A. and Nabergoj, R., A flow induced system with dry friction, *Chaos, Solitons Fractals* **7**, 353–365, 1996.

Tresser, C., About some theorems of L.P. Šilnikov, Ann. Inst. H. Poincaré **40**, 441–461, 1984.

Urabe, M., Galerkin's procedure for nonlinear periodic systems, *Arch. Ration. Mech. Anal.* **20**, 120–152, 1965.

Urabe, M. and Reiter, A., Numerical computation of nonlinear forced oscillations by Galerkin's procedure, *J. Math. Anal. Appl.* **14**, 107–140, 1966.

van der Burgh, A., On the asymptotic solutions of the differential equations of the elastic pendulum, *J. Méc.* **7**(4), 507–520, 1968.

van der Burgh, A.H.P., On the modelling of a continuous oscillator by oscillators with a finite number of degrees of freedom, in *Proceedings of the Third European Conference on Mathematics and Industry*, J. Manley et al., eds., Kluwer Academic Publishers and B.G. Teubner, Stuttgart, 1990, pp. 159–178.

van Gils, S.A., Krupa, M., and Langford, W.F., Hopf bifurcation with non-semisimple 1:1 resonance, *Nonlinearity* **3**, 825–850, 1990.

Verhulst, F., *Nonlinear Differential Equations and Dynamical Systems*, Springer-Verlag, New York, 1996.

Wiggins, S., *Introduction to Applied Nonlinear Dynamical Systems and Chaos*, Springer-Verlag, New York, 1990.

Win-Min, T., Sri Namachchivaya, N., and Bajaj, A.K., Non-linear dynamics of a shallow arch under periodic excitation—I. 1:2 internal resonance, *Int. J. Non-Linear Mech.* **29**, 349–366, 1994.

Yakubovich, V.A. and Starzhinskii, V.M., *Linear Differential Equations with Periodic Coefficients*, Wiley, New York, 1975, Vols. I and II.

Zholondek, K., On the versality of a family of symmetric vector fields in the plane, Math. USSR Sbornik, Vol. 48, 1984.

Index

absorber, 4, 45, 50
    tuned, 5
airplane wings, 5
amplitude-frequency diagram, 18
arches, vibrations of, 7
averaging, 93
averaging method, 17, 23, 27

bifurcation, 7
    Šilnikov, 52, 112, 117
    codimension-two, with eigenvalues $0, \pm i$, 28
    Hopf, 28, 44, 51, 55, 65, 117
    period-doubling, 21, 28, 44, 51
    pitchfork, 42, 116
    saddle-node, 28, 51, 117
bluff-body system, 90

centrifuge, 129
chaotic solution, 3, 8, 13, 28, 44, 52, 59, 65, 112
characteristic multipliers, 114
characteristic polynomial, 149
combination angle, 28
cross flow, 9, 31

domain of attraction, 4, 13, 23, 35, 101
dry friction, 5, 92, 99

Floquet theory, 114
Fourier series, 12, 54, 146
frequency
    interval, 3
    response curve, 19, 35, 80
    tuning, 4

Galerkin projection, 12, 31
galloping of high-voltage lines, 31
gyroscopic effect, 143

Hamiltonian system, 33, 121, 148
harmonic balance, 80, 93
Hill equations, 67
homoclinic cycle, 121
homoclinic orbit, 52
horseshoe, 128
hysteresis, 92, 130, 141
hysteresis jump, 52

instability
    domain, 16, 24, 34, 40, 42, 48, 69, 77, 81, 95, 134

Index

instability (*cont.*)
   interval, 3, 16, 27, 35
   threshhold, 17
invariant torus, 46, 113, 124

Lagrange equation, 46, 131
limit cycle, 44, 101

Mathieu equation, 16, 34, 48, 67, 76, 86, 95, 108, 129, 134
Melnikov method, 122

normal form, 112
normal mode, 1, 3

pendulum
   elastic, 1, 2
   parametrically excited, 1
   ring, 5
   rotating, 8, 13
phase-amplitude variables, 38
phase locking, 42
Poincaré map, 118
Poincaré–Lindstedt method, 20, 25, 28

quasi periodic, 3
quasi-periodic solution, 52, 61
quenching, 44, 47, 50, 58

resonance
   1:1, 148
   2:1, 1, 15, 45
   autoparametric, 7

internal, 7
parametric, 67
region, 13
Reynolds number, 91
Ritz method, 31
rotating periodic solution, 47, 56
rotor, 12
Routh–Hurwitz condition, 51

saturation, 4, 5, 21, 25
self-excitation
   Rayleigh type, 15, 43
   van der Pol-type, 37
separatrix, 101
spring
   leaf, 5, 8
   nonlinear, 9, 79
stability
   boundary curve, 19, 28, 35, 69, 95, 109
stiffness matrix, 67
strip theory, 89
structural stability, 146, 149
synchronisation effect, 28

upright equilibrium of a pendulum, 56

van der Pol equation, 113
versal unfolding, 148
vortex shedding, 91

whirling motion, 143
Whitney umbrella, 145